Low Substrate
Temperature Modeling Outlook
of Scaled n-MOSFET

Synthesis Lectures on Emerging Engineering Technologies

Synthesis Lectures on Emerging Engineering Technologies

Editor
Kris Iniewski, *CMOS ET*

Low Substrate
Temperature Modeling Outlook
of Scaled n-MOSFET

Low Substrate Temperature Modeling Outlook of Scaled n-MOSFET
Nabil Shovon Ashraf

ISBN: 978-3-031-00906-8   paperback
ISBN: 978-3-031-02034-6   ebook
ISBN: 978-3-031-00140-6   hardcover

DOI 10.1007/978-3-031-02034-6

A Publication in the Springer series
SYNTHESIS LECTURES ON EMERGING ENGINEERING TECHNOLOGIES

Lecture #10
Series Editor: Kris Iniewski, *CMOS ET*
Series ISSN
Print 2381-1412   Electronic 2381-1439

Low Substrate Temperature Modeling Outlook of Scaled n-MOSFET

Nabil Shovon Ashraf
North South University

SYNTHESIS LECTURES ON EMERGING ENGINEERING TECHNOLOGIES #10

ABSTRACT

Low substrate/lattice temperature ($<$ 300 K) operation of n-MOSFET has been effectively studied by device research and integration professionals in CMOS logic and analog products from the early 1970s. The author of this book previously composed an e-book [1] in this area where he and his co-authors performed original simulation and modeling work on MOSFET threshold voltage and demonstrated that through efficient manipulation of threshold voltage values at lower substrate temperatures, superior degrees of reduction of subthreshold and off-state leakage current can be implemented in high-density logic and microprocessor chips fabricated in a silicon die. In this book, the author explores other device parameters such as channel inversion carrier mobility and its characteristic evolution as temperature on the die varies from 100–300 K. Channel mobility affects both on-state drain current and subthreshold drain current and both drain current behaviors at lower temperatures have been modeled accurately and simulated for a 1 μm channel length n-MOSFET. In addition, subthreshold slope which is an indicator of how speedily the device drain current can be switched between near off current and maximum drain current is an important device attribute to model at lower operating substrate temperatures. This book is the first to illustrate the fact that a single subthreshold slope value which is generally reported in textbook plots and research articles, is erroneous and at lower gate voltage below inversion, subthreshold slope value exhibits a variation tendency on applied gate voltage below threshold, i.e., varying depletion layer and vertical field induced surface band bending variations at the MOSFET channel surface. The author also will critically review the state-of-the art effectiveness of certain device architectures presently prevalent in the semiconductor industry below 45 nm node from the perspectives of device physical analysis at lower substrate temperature operating conditions. The book concludes with an emphasis on modeling simulations, inviting the device professionals to meet the performance bottlenecks emanating from inceptives present at these lower temperatures of operation of today's 10 nm device architectures.

KEYWORDS

threshold voltage, substrate temperature, on-state drain current, subthreshold leakage current, bulk mobility, channel mobility, surface potential, inversion charge density, subthreshold slope, Tunnel FET, silicon nanowire FET, ferroelectric FET

Contents

Preface

Today's 10 nm device architectures are facing severe challenges and bottlenecks in terms of maintaining ITRS projection of higher I_{oon} or on-state drive current, lower I_{off} or off-state or subthreshold state leakage current, higher I_{oon}/I_{ooff}, lower subthreshold slope if possible below 300 K limit or 60 mV/decade, appreciable channel mobility to aid in transporting drift current, and most of adaptations of device architectures to accrue these projection goals are all being pursued at 300 K or room temperature device operation. The author of this book has previously written a book which introduced for the first time the concept of tailoring the substrate temperature between 100–300 K into the existing or exotic device architectures to control device threshold voltage to control leakage current reduction and drive current enhancement where substrate bias effects at these low temperatures are also accurately extracted for modeling purposes. Following the success of that book, the author has improvised, in the current book, a new outlook of device parameters controllability and adjustments using the same substrate temperature setting between 100 K and 300 K. Here the author outscores surface potential-based modeling equations and their concatenated integration to show higher drive current and lower off current at lower operating temperatures for conventional 1 μm channel length n-MOSFET device and values of these parameters can be suitably projected to keep up the similar I_{on}-I_{off} profile at various substrate temperatures in the vicinity of 10 nm technology node. Through analytical modeling equations the author has implemented novel modeling techniques and sequences to extract channel mobility as a function of vertical effective gate field adjustable at temperatures between 100 K and 300 K and the values can be standardized with experimental extractions of the channel mobility also at a reduced corresponding temperature operation than 300 K. Concept of gate voltage bias-dependent subthreshold slope was introduced as a potential new modeling finding in this book. The book discusses subthreshold slope factor as a function of gate voltage and for today's 10 nm devices built with Tunnel FET, nanowire FET, or ferroelectric FET, the single point subthreshold slope has to be supplanted by the average value of subthreshold slope in the vicinity of device threshold voltage, i.e., below threshold to close to threshold operation. Finally in a series of dedicated chapters the author for the first time critically analyzes the operational characteristics of tunnel-FET, silicon nanowire FET and ferroelectric negative capacitance FET and their comparative merits and disadvantages to be amenable to lower substrate temperature operation with high volume manufacturing in sight. The book will be ideal for the skilled perusers, research investigators, and industry professionals to gather key fundamental and well-reasoned device physical analysis to enhance their assessment of the potential benefit of lower substrate temperature operation from the perspective of efficient analytical and computational model development and studying in-depth the novel device architectures deployed by the industry at the 10 nm technology node era from their adaptability considerations to lower substrate temperature operation whether these proven benefits can be actually realized when these devices are available to consumers for various microelectronic and nanoelectronic applications.

Nabil Shovon Ashraf
June 2018

Acknowledgments

The author has been assisted at various stages during the composition of this book by the proper configuration of simulation outcomes exhibited and analyzed in this book with regard to scaled device performance when low substrate temperature operation of n-MOSFET is considered. The author is grateful to the diligent time of research hours devoted by the students Md. Nuruzzaman, Mahabub Ur Rahman, Somana Ali, K.M. Shihab Hossain, and Saifullah Basher Shohel of the Department of Electrical and Computer Engineering of North South University, Dhaka, Bangladesh.

Nabil Shovon Ashraf
June 2018

CHAPTER 1

Introduction

Due to the demand of high packing density of micro-manufactured chips on a silicon die to support a host of performance applications of systems such as high-performance Tera Hz microprocessor, systems-on-chip, 3D integration and packaging, RF and analog application-specific integrated circuits, digital signal processing core architecture systems, and other highly evolving yet miniaturized systems, the transistor device size is getting down to almost single nm dimension regime. Under such scaling impetus on device size miniaturization, the most a transistor suffers is from its own intrinsic performance capability limits. Over the last 10–12 years, we have witnessed a myriad of architectural breakthroughs on device structure and technology evolution [20–40], [42–70], [74–112] that have governed the required feature size reduction of a transistor up to 10 nm and for some time now we also witnessed a plateau at this node where industry and institutional research professionals are hesitant to forecast what future architecture will assume once the feature size of a transistor enters 7 nm and below. In a previous book [1], also written by the current author, it has been brought to the attention of industry manufacturers and device research professionals toward a novel prospect of sustaining the scaling scenario at every technology node below 10 nm in accordance with Moore's Law-based scaling by masterfully integrating lower substrate temperature with the developed device architecture pursuant to that technology node of scaling. In fact, with all the ingenuity and brainstormed architectural breakthroughs that have provided us the transistor device outlook at 10 nm and below, if the die temperature regulation capability controlled by a temperature-setting digital logic unit as described in the previous book [1] is enabled at various sites of the die partitioned by areas of different logic units including interconnects and peripherals, the overall performance of the gazillions of transistors in a silicon die can be improved to a considerable proportion as the operation of lower substrate or lattice temperature not too below 300 K will ensure intrinsic performance benefits and leverage its limits that is set technology node scaling and also additionally prolong device operational lifespan by improving transistor reliability which is becoming the principal bottleneck to scaled device operation in the vicinity of a 10 nm regime.

Since the occurrence of 22 nm and below, today's scaled transistors are subjected to reliability failures emanating from the presence of interface and oxide traps in the MOSFET. A very minute number of these traps originate from manufacturing processing temperature-stressing cycles and a few other oxide traps result from sustained direct oxide field from the gate voltage that both the active and idle transistors experience during the operational process of the electronic systems. The interface traps are resulting from the inability of attaining seamless interface between silicon and the silicon dioxide overlaid film. These traps randomly trap and emit conduction carriers of an n-MOSFET when the device is fully ON giving rise to various forms of noisy profiles such as low-frequency or 1/f noise and random telegraph noise (RTN) which is present both at low and high frequency in highly scaled device structures with extremely scaled gate width. The widespread research and experimental demonstration of various facets of RTN needs to be taken seriously as RTN effect cannot be minimized by device architectural improvements; rather, an exhaustive understanding of the nature of the trap's interaction with conduction carriers in the vicinity of a trap's capture zone reveals the fact that the capture and emission process by traps does depend on substrate or lattice temperature and lowering the substrate or lattice temperatures less than 300 K or room temperature can control both trap's capture and emission kinetics eventually leading to RTN effect reduction and invigorating the transistor's reliable performance. These added benefits of lower substrate temperature operation simply provide tunability to architectural mutations that the researchers have successfully attained through high-volume batch manufacturing of structures such as fully depleted silicon-on-transistor (FD-SOI), silicon FinFET, Tunnel FET, and silicon nanowire high-k FET, leading to nanowire Tunnel FET. From our knowledge of experimentally demonstrated MOSFET device structures in research labs and manufacturing platforms targeting the performance benefits that have been realized from drain current enhancement, subthreshold drain or leakage current reduction, carrier mobility enhancement and subthreshold slope reduction close to 60 mV/decade or below, additional integration of lower substrate temperature with the base silicon material on which these devices have been lithographically defined can provide supplemental efficiency on drain current enhancement, subthreshold drain or leakage current reduction, further mobility enhancement and further reduction of subthreshold slope already achieved by scaled device architecture at a particular node particularly at 10 nm or below.

The book is organized as follows. Chapter 2 provides an overview of scaled MOSFET evolution beyond 45 nm and with respect to key device parametric modeling and strategic improvements accrued by different scaled architectures. Seminal articles from reputed journal and conference proceedings [20–40], [42–70], [74–112] will be brought to the attention of the perusers of this book from the perspective of detailed modeling of device physical parameters and how these parameters have been modeled to exhibit superior device performance projected from scaled device set-up and ITRS roadmap. The first book [1] composed by the author of this current book had included critical analysis of the modeling outlook of threshold voltage improvement and stability with regard to key industry device architectures. Other than threshold voltage, one of the most important device parameters that is being intensely pursued and researched for modeling purpose is the transistor channel mobility. Transistor carrier or channel mobility directly impacts MOSFET drive current or ON current, channel drain to source conductance R_{ch}, gate to channel transconductance and subthreshold drain current or the off-state leakage current. In fact, in the vision of highly dense 3D stacked or integrated architectures that we are witnessing today along with source and drain parasitic and source-drain extension and junction-related spreading resistances, the urgent need is to maximize the channel carrier mobility through device modeling and continued architectural modifications. Toward that goal we are already achieving the limits of channel carrier mobility, even utilizing longitudinal, transverse directions with regard to current carrying plane by suitably engineering the energy (E)-wavevector (k) band profile of valence band and conduction band of pure silicon, strained silicon structure, III-V heterojunction MOSFET, silicon-germanium MOSFET, modulation doped HEMT where the effective conductivity mass of silicon or other silicon heterojunction material is adjusted to the extent that near valence band edge and conduction band edge, the effective mass, is tuned to be lowered to ideally boost the device carrier mobility. The limits to which we can boost the channel mobility is growing fast due to high packing density and mostly 3D integration, contact and transmission interconnect-related parasitic resistances and generations of various coupling parasitic capacitances due to a mix of aspect ratios of width and height of multiple barrier layers along with dielectric isolation layers are gradually slowing the intrinsic speed of MOSFET where the very limits of modeled improvement of mobility is poignantly encountered. Therefore, the review analysis will highlight this aspect of the most striking aspect and demonstration of mobility improvement and explain why lower substrate temperature operation will keep up the mobility improvement by first setting the peak mobility based on phonon scattering-related transport where substrate temperature has most direct bearing. By setting the peak mobility at a sufficiently high value and having a detailed analysis of all the parasitic resistances and capacitance-related circuit propagation delay, it can be safe to say that the limits of mobility improvements by the present day tactics employed at room temperature of 300 K can be extended to a higher threshold and maintain superior circuit speed required by high-performance Tera Hz microprocessors. Chapter 2 will also highlight the modeled approaches referenced by these articles on drive-current, subthreshold leakage current and subthreshold slope improvements. The extent to which carrier mobility impacts the suitable control of drive-current as per ITRS roadmap and subthreshold leakage current will be probed with prospective contributions that can be suitably imparted from lowered substrate temperature operation of n-MOSFET device structures or its variant architectures.

Chapter 3 will document the original simulation results on subthreshold drain current at temperatures below 300 K from a 1 μm gate length n-channel MOSFET reference device. The chapter will also provide drive ON current simulation results at temperatures below 300 K. Detailed modeling equations with respect to different device-centric parameters that directly or implicitly connected to subthreshold leakage current calculations and on-state drive current calculations will be outlined. The modeling profiles of device-centric parameters such as threshold voltage and mobility will be developed with regards to various substrate temperatures. One of the key aspects of these accurately developed modeling equations for subthreshold drain current and on-state drain current is the ability to provide continuity in the drain current plot as a function of gate voltage biases particularly in the subthreshold of weak to moderate inversion regime. The simulation results will be shown to reveal important derivations about the lowered substrate temperature operational advantages for both subthreshold drain current and on-state drive current. Significance of the findings of these simulated outcomes of subthreshold drain current and drive ON-current on today's scaled device architecture scenario down to 10 nm will be addressed and discussed.

One of the fundamental and central figure-of-merits for improving circuit propagation delay and minimizing switching time is the channel mobility when the MOSFET is driven into ON position (linear mode or saturation mode). While the scaling of MOSFET continued up until today, the plethora of research articles documenting the research and improvisation on MOSFET channel mobility essentially comes to the very conclusion that channel mobility is reaching its plateau of operational magnitude due to constraints imposed by scaled miniature device size where substantially reduced gate width impacts the drastic reduction of channel mobility. The gate architecture with stacked gate and spacer oxide and 3-D integration also affects the channel mobility negatively. Besides, previously with scaling of MOSFET channel length, the substrate doping needed to be very high reaching several $10^{18}/cm^3$ in order to avoid the source and drain depletion widths from encroaching and margin within any part of the channel formed between source and drain regions. As a result, even with vertical gate voltage of nominal value, severe surface roughness scattering took place at the surface of the channel where the inversion charge centroid resides. Surface roughness scattering which asymptotically decreases the channel mobility by a factor of -2 precipitates the value of channel mobility at all temperatures including lowered temperature operation below 300 K. Therefore, when the concept of double gate MOSFET and FinFET MOSFET were posed in research articles, the substrate was doped intrinsically both in double-gate MOSFET and FinFET so that interface roughness scattering can be avoided even when the vertical field is enhanced due to ultra-low dielectric thickness after gate voltage scaling is put forth. Intrinsic doping additionally allows one to set the peak mobility determined by phonon scattering, which is substrate or lattice temperature related, to a considerably higher value compared to the case when substrate doping is in the range of $10^{16}/cm^3$–$10^{18}/cm^3$. In the case of double-gate MOSFET where the substrate doping is near intrinsic, a unique feature called volume inversion of the channel takes place which further boosts the channel mobility by screening the ionized impurity-related scattering. In Chapter 4, the author will systematically address the ingenious methods that have been adopted so far to control the channel mobility to its higher limit in order to benefit for femtosecond or attosecond switching time required for today's high-performance microprocessor CPU units. I have already insinuated in this section that even with design of mobility enhancement factors, such as including strain engineering in the channel, embedding stressors in the channel and also in source and drain regions, causing the direction of current flow in transverse direction with regard to substrate orientation determined by wafer flat, etc., the principal bottleneck exists on having been able to only achieve mobility of the order of 100 cm^2/V-s in the vicinity of 10 nm gate length node. In order to push the maximum channel mobility we are witnessing today to at least 1.15 times, the reference mobility value that comes out of fabricated devices on the die, lowered substrate temperature operation in the range 150–250 K can be of great aid owing to the fact that the peak value of the channel mobility which is determined by substrate temperature will be at a proportionately higher level compared to room temperature operation of MOSFET. In this context, the chapter provides original simulation results for channel mobility of electrons as a function of temperature for 100–500 K for different substrate dopings in the levels of $10^{15}/cm^3$–$10^{19}/cm^3$. Modeling equations and encompassing device physical attributes will be documented and addressed to critically analyze the plot. The modeling equations used to generate the channel mobility do not include any empirical fitting parameters that were reported by a key researcher and his group with a highly visible journal article. The simulation results also compares well with journal article references provided by N. D. Arora and John Hauser [113] who have also fundamentally contributed to adopting these device-centric models in some of the industry-standard TCAD models for simulating MOSFET channel mobility. Simulation results are also provided for vertical gate field-induced channel mobility as a function of vertical gate field for substrate temperatures of 300 K and 200 K. The simulation results were extracted by truly developing analytical models for channel mobility including substrate temperature as a parameter. The simulation outcomes compares well with the highly cited and popular Shinichi-Takagi mobility model and plots enunciated in [71] although the accuracy of the modeled values of mobility in Takagi's paper may be compromised by added simulation time complexity. A critical analysis of this reference paper will be underscored in Chapter 4 with added discussion on the benefits, simpler characteristics of the new modeling equations, developed by the current author, and used to generate channel mobility vs. gate field plots at $T = 300$ K and $T = 200$ K for substrate doping of $10^{16}/cm^3$ for a channel length of 1 μm n-MOSFET.

Another important device benchmark metric for evaluating drain current characteristics at the lower gate voltage biases where off-state leakage current and subthreshold current are critical is the subthreshold slope. Steeper subthreshold slope is (gate voltage change for decade change in drain current of a MOSFET) is important for swiftly turning on and off the MOSFET device. Chapter 5 will be solely concentrated on documenting the important and pivotal significance of MOSFET subthreshold slope and the distinctive outlook on how the room-temperature thermal activation limit of the diffusion of carriers over gate-source voltage barrier presently 60 mV/decade can be reduced even further by lowering the substrate temperature close to 200 K. The modeling parameters that contribute to subthreshold slope when the thermal activation and diffusion process is considered will be highlighted with their associated substrate temperature-related dependence fully analyzed. Simulation results of subthreshold slope factor vs. gate voltage for 1 μm MOSFET for different substrate temperatures 100–500 K will be presented in this chapter. For the first time, detailed gate vóltage-dependent surface potential modeling-based analysis extended from 100–500 K show that the textbook reference of constant subthreshold slope when the device is at subthreshold region is slightly erroneous. Rather based on the gate voltage bias in the vicinity of subthreshold region, the subthreshold slope factor n from which the subthreshold slope is computed shows slight variation as the gate voltage is on the higher end of the subthreshold bias and exhibits good amount of variation as the gate voltage is further reduced closer to 0 V or for a few 20 mV. A constructive and illustrative table is thus presented in the chapter that computes the gate voltage bias-dependent subthreshold slope factor for a temperatur range of of 100–500 K for a given set of gate voltage biases. Extraction of this subthreshold slope variation is important as for scaled devices when device architectural improvements in the form of gate-all-around nanowire and Tunnel FETs are devised in the vicinity of gate length of 20 nm or less, even 5–10 mV/decade subthreshold slope deviation from expected device-level average values of the fabricated die in the integrated circuits can provide substantial variation in the intra die and die-to-die leakage current estimation particularly during off-state and subthreshold state. Variation of subthreshold slope also imparts dispersion in the rate of turn on/off of the device as a switch and can impact all digital circuit core blocks of a high-performance microprocessor or core ASIC microcontroller chips including DSP processors and non-volatile memory (NVM) blocks. Properly estimating the variation of subthreshold slope as a function of gate voltage biased in the region of fully-off to subthreshold region at different substrate temperatures of operation of MOSFET is highly informational and the table and simulation plots provided in this chapter will surely be interests to the device engineering professionals in the industry and academia who are constantly engaged in scaled MOSFET evolution and performance scenario.

Chapter 6 provides important review and critical analysis-based contents for industry standard device architectures at the scaled node of 10 nm, namely Tunnel FET, gate-all-around nanowire FET, and ferroelectric negative capacitance FET. The capabilities and device performance assessment as per ITRS projection for scaled nodes has been critically assessed for these device architectures in this chapter. In addition to the perspective of lower substrate temperature operation, the operational benefits and bottlenecks of these devices are also critically analyzed.

Chapter 7 will further provide augmented analysis considering the resulted simulation outputs of subthreshold drain current, mobility, and subthreshold slope from the importance of lower substrate temperature operation. Since the goal is increasing device performance as we approach 10 nm gate length or lower node of operation, the proper assessment of lower temperature operational benefits accrued from reduced subthreshold drain current, increased channel mobility and decreased subthreshold slope less than 60 mV/decade will be weighed upon and discussed. From this chapter, device physicists and device engineering professionals should be able to detect possible clues to extend the bottleneck we are observing at scaled node of operation in the vicinity of 10 nm on subthreshold drain current limits, maximum possible on-current limits, reduction of subthreshold slope beyond ideal thermal energy limit at room temperature, and limits that are observed on manufactured MOSFET device on channel mobility.

In today's manufactured die in silicon wafer, die-to-die and within-die parametric variability exists due to process-induced traps and traps originating from voltage stress or gate and drain field-induced stresses during sustained operational condition of MOSFET. These traps lie at the interface between silicon dioxide and silicon channel surface and can be randomly positioned from source end to the drain end of a MOSFET. In addition, traps in the oxide can be generated very close to the silicon dioxide-silicon interface layer also known as fast traps and can be progressively situated toward the middle of oxide thickness also known as shallow border level traps. Stochastic capture of channel electrons or filling of the traps and emission of channel electrons from the filled traps or emptying the traps can give rise to a variety of reliability related device failure or parametric drifts events. Three such effects that are currently proving to be major show stoppers for reliable operations of atomically sized MOSFET are (a) random telegraph noise or signal (RTN/RTS), (b) negative bias temperature instability (NBTI), and (c) positive bias temperature instability (PBTI). On top of this, as the channel dimensions shrink both in length and width, the actual number of dopant electrons becomes much less numerous and sparsely distributed in the depletion region beneath the channel resulting in random dopant fluctuations (RDF) in the depletion region vertically from semiconductor surface and also laterally from source side junction to drain side junction. The random dopant fluctuations get factored into the reliability phenomena RTN, NBTI, and PBTI and cause substantial drift in threshold voltage and degradation of channel trans-conductance from fluctuations in channel conductance resulting from inversion carrier fluctuations and additionally by fluctuations in channel mobility resulting from scattering related fluctuations of inversion carriers from traps located near or at the SiO_2-silicon channel interface. Finally, Chapter 7 concludes with the preeminent importance of modeling equation development for every kind of device architecture benchmarking at different substrate temperatures less than 300 K. When the precise and systematic derivation of modeling equations for different device parameters are generated for each state-of-the art device architecture, modeling library can be developed to tailor the device parametric centered performance as a function of substrate temperatures. Then utilizing the substrate temperature regulation and control scheme in the fabricated multifunctional die as laid out in previously published books [1, 72, 73], the modeling equation-based libraries thus generated can be of potent aid to the device manufacturing professionals to engineer successful performance of scaled n-MOSFET even in the midst of a reduced window of operational benefits at the very end of Moore's Law.

CHAPTER 2

Historical Perspectives of Scaled MOSFET Evolution

From the exhaustive and pinpoint survey of the literature demonstrating scaled MOSFET evolution in last 10–15 years [1–19], [20–41], [42–71], [74–112], the promising device architectural configurations as determined by ease of manufacturability and device parameter-centric performance benefits have been extracted to be double-gate MOSFET, FDSOI double-gate MOSFET, FinFET and its variant such as Tri-Gate, Tunnel FET and gate-all-around non-planar architectures including nanowire FinFET and Tunnel FET. The wide plethora literature concentrates on detailed device and process modeling of these devices with the goal of (i) on-state current magnitude boost, (ii) leakage current reduction, (iii) superior I_{on}/I_{off} ratio, (iv) subthreshold slope amenability close to room temperature (300 K) limit or 60 mV/decade, and (v) mobility improvement by employing additional intrinsic parameter modifying deployments such as stress and strain and wafer orientations. At present, all these adapted incentives to device architectures are performed at 300 K as, although evidence is now on hand about additional performance improvements at temperatures around 77 K, the apparent prohibition to alter substrate temperature stems from the observation that the cost of cooling the chip by die interfaced cooling instruments is much too high losing favorability on the basis of sustained cost reduction economics taking the price of a single chip as available from market. Some literature surveys in the past have concentrated on device simulations and experimental demonstrations revealing that at lower substrate temperatures the intended aim of engendering higher device on-current as the technology scales on gate or channel length of MOSFET is falling short of the mark. This is due to the fact that one of the key parameters that determines on-current rise is gate overdrive $(V_{gs} - V_T)$ where V_{gs} is gate-to-source applied voltage and V_T is the device threshold voltage for a particular scaled channel length n-MOSFET device. As the substrate temperature is reduced, V_T is enhanced from its room temperature counterpart and in fact as the temperature is more reduced from 300 K reference value, the shift of threshold voltage in the ascending direction can be very high. Considering that with constant field scaling as currently employed for scaled MOSFET evolution, voltage scaling proportional to channel length scaling is required which makes available V_{gs} (V) lower with scaling evolution of each technology generation. Consequently, the voltage overdrive $(V_{gs} - V_T)$ can be very low indicating diminishing amount of inversion charge density with scaled channel length if we do employ a lower substrate temperature to fabricate these reference device structures. In addition, mobility enhancement that comes from the intrinsic nature of silicon carriers at lower substrate temperature mostly from peak mobility value as set by phonon-related mobility might not be doubling even when the substrate temperatures are near the vicinity of 100 K. It is due to the observation that the concept of channel mobility hinges on the surface related mobility and owing to the reduction of gate oxide already in the range of less than 2 nm, high vertical gate to body field exists at the surface shifting the operational region of the mobility as a function of gate vertical field toward the interface roughness scattering and as we have mentioned in the first book composed by present author [1], interface roughness scattering highly conspicuous at reduced substrate temperature operation of n-MOSFET does not let the device professional to achieve the targeted mobility enhancing peak value originating from temperature dependent phonon related mobility term. Also, high vertical field due to greatly reduced oxide thickness reduction draws the carriers near the surface, and with lesser concentration of inversion charge density the carriers have no strongly correlated momentum as they transport along from source to drain of a MOSFET. This is the second reason that the transport feature in this way degrades the channel mobility from its initially projected higher intrinsic carrier mobility as determined from reduced substrate temperature setting. But just because on-current enhancement is not sufficiently achievable by our targeted reduced substrate temperature operation does not mean that the added cost of making such lower substrate temperature integration to device operation would not be viable from a production standpoint with consumer satisfaction. If we take a second look at today's highly dense and integrated VLSI circuits, the nearly constant trend of achievable device on-current per unit gate width with scaling is managed by integrating more die on a functional chip. As a result, the cumulative nature of multimillion or billions of die will always bring the aggregated device-on-current commensurately with scaled MOSFET performance projection. Since lower substrate temperature operation does not aggravate the on-current value from its 300 K reference value even though the on-current at lower temperature has only slight increments, this is still preferable and advantageous from a high-density VLSI integration perspective, as described above. On the other hand, today's high-performance microprocessor units with their billions of die in the chip form suffer from increasing thermal dissipation or heating per device or chip. This is due to the fact that with the scaled MOSFET channel length there is an exponential rise in subthreshold leakage current per die and with billions of die integrated in a microprocessor chip, the growing heat capacity of the die is already in the range of 1 W/cm^2—a very alarming and stupendous number already marked by device professionals when the die is fabricated for 300 K operation. When the substrate temperature is reduced in the range 100–200 K, the subthreshold leakage current for a die is reduced orders of magnitude making the total subthreshold leakage current of the system fractions smaller compared to its 300 K counterpart. As a result, the heat dissipation of the microprocessor slows down for long-term operation of the device when a significant portion of the system draws off-state leakage current due to its idle state of operation. The most significant advantage from reduced substrate temperature operation thus comes from reduced power dissipation in the die from a substantially reduced leakage current at these operated temperatures. In addition, another key performance benchmark metric I_{on}/I_{off} also benefits from lower substrate temperature operation owing to the orders of magnitude reduction of I_{off} or off-state leakage current. The steeper subthreshold slope less than 60 mV/decade is more conveniently achievable at reduced substrate temperatures for all the device architectures mentioned in the beginning of this chapter. In fact, these devices already use multi-gate structure to improve gate integrity to control subthreshold slope lower than room temperature value. On top of that, the direct reduction of substrate temperature translates into proportional reduction of the subthreshold slope factor or subthreshold slope value and these observations finally let us to the conclusion that operating the device at reduced substrate temperature is superiorly sustainable and feasible from device scaling and manufacturability concerns pending the fact that the only metric where the benefit of device performance may not be significant is device-on-current as prescribed by device scaling requirements and the authors have put forth alternative ways to still preserve this needed benefit while adopting a lower substrate temperature device operation scheme.

The currently deployed device architectures that have accordingly evolved as per Moore's law-based scaling and ITRS roadmap are Fully FDSOI leading to ET-SOI (Extremely Thin Silicon-on-Insulator), Fin-FET with the concept of gate dimension arranged thin fin shape height and width, and Tunnel FET where electron transport is controlled by electron tunneling from $p+$ (source) to $n+$ (base) forming abrupt junction with extremely narrow tunneling width and subsequently drifting through i (intrinsic layer) and collected by reverse bias field existing at $n+$ (drain). As the device dimension reaches 10 nm, gate-all-around (GAA) nanowire architectures for a double-gate, silicon-on-insulator and Tunnel FET are adopted. In [1], electrostatics from device physical analysis was enumerated for double-gate device architectures from the context of reduced substrate temperature operation. Reduced substrate temperature operation benefits from the generally accepted requirement for double-gate MOSFET in terms of intrinsic channel doping for volume inversion and as it has been described in [1], intrinsic doping sets the highest possible channel mobility available at reduced substrate temperature by phonon scattering which gets substantially reduced when the doping is intrinsic. The higher achievable channel mobility at reduced substrate temperature operation less than 300 K is almost 2–2.5 times using the intrinsic doping of the body of the double-gate MOSFET when we are already experiencing mobility threshold currently devisable from device architectures in the 100–150 cm²/V-s range and this is an excellent breakaway denouement considering the stringent ITRS roadmap based device performance benchmarks required at every scaled node.

The ET-SOI architecture is a derivative from FDSOI device where the body is fabricated extremely thin for better electrostatics from gate control and lesser leakage current flow from source to drain through the substrate underneath the channel owing to the thinness of the body. ET-SOI MOSFET architectures have been manufactured by the IBM research group with excellent subthreshold slope and drain-induced-barrier-lowering (DIBL) values. ET-SOI MOSFET can benefit from reduced substrate temperature operation in several ways. First, the channel doping can be reduced as double gate for intrinsic doping and this will result in higher carrier mobility. The widening of the depletion region in the thin body due to intrinsic doping might result in a slightly increased source-to-drain subthreshold leakage current but due to the moderately increased or tuned threshold voltage at a reduced substrate temperature, the leakage current will still be reduced. The subthreshold slope always benefits from planar MOSFET-like architectures when the channel doping is near intrinsic region enabling the reduction of depletion capacitance from channel interface to body that appears as a factor controlling the subthreshold slope factor. The other benefit comes from the body potential of the extremely thin SOI MOSFET which modulates the threshold voltage of the device more strongly than thin FDSOI MOSFET. By properly selecting the substrate temperature, the back-gate body bias of the extremely thin SOI MOSFET can be engineered in a programmable loop to set the required threshold voltage of devices on a die which should be on and fast and devices on a die which should be either idle or off or should be slower in response to byte transmission. This benefit is also one of the potent feature of double-gate FDSOI MOSFET where back-body bias is always controlled to monitor and assign the threshold voltage of various devices at a particular location on the die to control device on-current and off-current (leakage current) and power consumption in a core device functional block on the die due to aggregated leakage current.

At this juncture, the author of this book would like to draw the readers' attention to some important device performance scenarios where particularly low-temperature operation of MOSFET provides critical advantage. As the scaling continued early 2000 era, the channel lengths of the MOSFET were in the vicinity of sub-quarter micron and then below 0.1 μm or 100 nm. The substrate doping required to properly isolate the source-substrate and drain-substrate junction extensions was in the vicinity of more than 5×10^{18} cm^{-3} which at room temperature 300 K can be considered to be degenerate doping condition. In fact, around 90 nm when auxiliary structural adaptations to conventional MOSFET structure is still preserved in scaling evolution, the substrate doping can be in the vicinity $> 8 \times 10^{18}$ cm^{-3}. Why is the degenerate doping deleterious from a device performance aspect will be explained now.

The band-gap of a semiconductor, particularly silicon, decreases with high degenerate doping. The effective masses of conduction and valence band show a high degree of non-parabolicity effect in the E (energy)-k (wavevector) curvature resulting in enhancement of mass values. As a result of band gap narrowing and effective mass increase of electrons and holes, the intrinsic carrier concentration increases sharply from its already defined room temperature value. This effect has a direct consequence on the value of the bulk potential of n-MOSFET where acceptor-type doping is considered. As a result of intrinsic carrier concentration increase up to a few 10^{11}/cm^3, the threshold voltage will be partly increased by $(N_A)^{0.5}$ factor in the bulk charge term and will be partly decreased by $2\varphi_B$ term where φ_B is the bulk potential decremented by the logarithmic factor where n_i, intrinsic carrier concentration is inversely included. Here, we are still limiting the maximum ionization to be 100% of dopant concentration although when the Fermi energy level is closer to conduction and valence band within $3\,kT$ band where kT

is the thermal energy in eV at room temperature 300 K, the free carrier concentration will increase. The flat band voltage is also less negative as $(E_i - E_F)$ term-defining bulk potential has decreased noting E_i will be further driven toward valence band E_V owing to the fact that both effective masses of electron and hole have increased. Overall, the contribution from decrements of bulk potential will not be substantial and also increment of flat-band term will have a minute effect on overall threshold voltage increase by larger addition from $(N_A)^{0.5}$ factor in the bulk charge term. In addition, heavy degenerate doping might lead to quantum confinement of carriers consequent to additional contribution to threshold voltage increase resulting in reduction of gate overdrive during saturation condition of the MOSFET and reducing device circuit speed and transconductance. It is also worth noting that intrinsic carrier concentration due to degenerate substrate doping condition will translate into an increase of subthreshold current and direct source to drain leakage current which is proportional to $(n_i)^2$ where n_i denotes value of intrinsic carrier concentration. We also observe that as a direct bearing of many-body effects resulting from significant doping condition of the substrate, device equilibrium is disturbed and equilibrium carrier-physics related all mathematical derivations will be inaccurate. Also, the carrier inversion mobility will be significantly affected by carrier-to-carrier scattering and also increase surface roughness scattering. Due to degenerate doping, the source to substrate doping profile will be very steep and as a result band-to-band tunneling leakage in addition to direct source-to-substrate to drain leakage current and subthreshold current will be dominant. With regard to quantum mechanical (QM) confinement, a condition that exists at the Si-SiO$_2$ interface when degenerate doping condition is encountered due to aggressive scaling requirement, it is true that quantum confinement adds an additional upshift in the energy band gap leading to some proportional decrease of intrinsic carrier concentration and this will additionally counter the original decrease of bulk potential from pure degeneracy without QM effect. Also, the actual increase of n_i as indicative from pure degenerate condition of substrate doping will be compromised due to QM shifted band gap increase-related n_i reduction. But the overall impact of various leakage current-contributing components due to degenerate substrate doping condition will be mostly unaltered by incorporating the QM effect, a consequence arising from heavy substrate doping. When the substrate temperature is reduced from $T = 300$ K to $T = 100$ K, all the present day device architectures by dint of surrounding the device with gate-stack on top, bottom and side faces for improving gate-controlled channel potential integrity have been tailored to employ near intrinsic-level substrate doping. The feasibility of employing intrinsic doping when surrounding gate architectures are considered and aided by the condition of reduced substrate temperature operation readily removes the burden from device physicists and engineers to employ doping conditions closer to the degenerate level and avoid all the deleterious device performance factors detailed above.

CHAPTER 3

Simulation Results of On-State Drain Current and Subthreshold Drain Current at Substrate Temperatures Below 300 K

At the limits of aggressive device scaling being imposed by technology nodes on device architecture evolution of MOSFET, due to giga scale integration density, the off-state or idle-state leakage power dissipation in a chip is exceeding more than $10\,W/cm^2$ and causing severe thermal management issues of proper heat dissipation and regulation on-chip. Device engineers therefore have focused on reducing leakage power by properly attenuating the subthreshold leakage current or off-state leakage current of MOSFET at every technology node. The major reduction precursors of subthreshold leakage current are threshold voltage and intrinsic carrier concentration. By operating every technology node of architecturally reconstructed MOSFET at reduced substrate temperatures close to 100–300 K, the threshold voltage can be suitably increased at a lower temperature and intrinsic carrier concentration can be decreased orders of magnitude below its room temperature value. Since the subthreshold drain current is exponentially dependent on threshold voltage and it is also dependent on a square of intrinsic carrier concentration, we can see that a slight increase of threshold voltage can exponentially reduce the subthreshold drain current magnitude to a handsome proportion and, additionally, the squared reduction of intrinsic carrier concentration can reduce the subthreshold leakage current in a more direct way when all the device architectures employed for technology nodes up to 10 nm are tailored to operate at reduced substrate temperatures less than 300 K. One very potent revelation that we point out here is that from a long-channel MOSFET operational perspective, the subthreshold condition actually means operating at a gate voltage to some extent lower than threshold voltage. The concomitant reduction in the leakage current that we pointed out due to squared reduction of intrinsic carrier concentration is feasible only when the gate voltage applied is not of the order of threshold voltage of the MOSFET but a certain value lower than it. Then it becomes obvious that long-channel subthreshold current modeling actually generates highly decremented leakage current value as substrate temperature is lowered. As the scaling evolution continuously scales the channel length, it also scales the power supply voltage or voltage to be applied on MOSFET gate terminal. As a result, for highly scaled MOSFET device such as channel length being of the order of 10 nm, the actual gate voltage that determines subthreshold condition for this MOS-FET cannot be lowered more than a few thermal voltages below the scaled threshold voltage value at that particular node. As a consequence, important adjustment takes place in the sub-threshold current modeling equation and compared to intrinsic carrier concentration squared term-based reduction, surface potential incorporating exponential term based reduction is only moderate at lower substrate temperature. Combined effects of these two terms allow reduction of leakage current with reduction in substrate temperature when channel length is at the short channel regime but the scaled supply is still within 0.5–0.7 V.

With the goal of accounting the impact of lower substrate temperatures less than 300 K on the subthreshold leakage current in an n-MOSFET, the authors have simulated the long-channel n-MOSFET with gate length $L = 1$ μm, nominal oxide thickness of 80 nm, and nomi-nal substrate acceptor doping of 10^{16}/cm^3. The threshold voltage V_T for this n-MOSFET device at 300 K is 1.12 V and the modeling equations utilized to compute this value is systematically described in [1]. Incomplete ionization effect properly accounted for substrate temperatures in the vicinity of 100–200 K with its doping dependent round off of the ionization factor. We now chronologically enumerate the modeling equations that were developed to extract the drain current-drain voltage characteristics for this 1 μm MOSFET at different reduced substrate tem-peratures lower than 300 K. From the drain bias and gate bias-dependent drain current values of these simulated outcomes figures at the technology node of $L = 1$ μm and $t_{ox} = 80$ nm, proper vertical field-based scaling and adapted version of it as available in literature can be used to project the corresponding drain current value for a scaled drain bias voltage and gate bias voltage as is required in the vicinity of 10 nm node. Although short-channel effects based mod-eling adaptations are not included in these modeling derivations, particularly when lower than room temperature operation of a device is considered, these figures can still be used as guiding references to set the required drain current value at a technology node reaching 10 nm with the required drain and gate voltages of operation with a moderate variability window across the mean nominal value determined by process variability due to substrate temperature oper-ation at lower than 300 K and other short-channel device architecture-related variability such as drain-induced barrier lowering, quantum confinement, carrier degeneracy although intrinsic device doping is currently employed at reduced technology node architectures, source-to-drain punchthrough current, some form of band-to-band tunneling current which is ideally not ema-nating from device doping of the substrate but more from the intense vertical gate field resulting from oxide thickness below 1.5 nm all extracted at the operating substrate temperatures. As is detailed [1], on-state drain current will be mostly impacted by threshold voltage at a selected substrate temperature and gate-field-dependent mobility of inversion carriers at a selected sub-strate temperature at every technology node. From the modeling equations a better projection of mobility values and threshold voltage values at a scaled technology node for these reduced substrate temperatures can be deduced and with a slight variation caused by short channel factors at different reduced substrate temperatures, the nominal drain current values thus scaled to evolve at projected technology node with its minute variation can still be accurately extracted from the simulated outcomes detailed in this chapter for on-state drain current values at different substrate temperatures for long-channel n-MOSFET of channel length 1 μm.

3.1   MODELING EQUATIONS TO DERIVE ON-STATE DRAIN CURRENT AS A FUNCTION OF DRAIN VOLTAGE FOR DIFFERENT GATE VOLTAGES OPERATED AT REDUCED SUBSTRATE TEMPERATURES BELOW 300 K

For an n-channel enhancement mode MOSFET, one of the most important performance figures-of-merit is the on-current drive value of the MOSFET as the technology node evolves modifying its channel length and gate width value. From a low-temperature modeling perspective, it is essential for device modeling engineers and professionals to precisely estimate this drain current value at a gate voltage close to threshold voltage value exceeding it and then at a higher gate voltage value within the maximum voltage scaling threshold with first a low drain voltage and then at a higher drain voltage. From room temperature 300 K this drive current must be calculated at lower than room temperature substrate temperatures and sometimes higher than room temperature, when high temperature electronic applications are needed, with every technology node impacting the threshold voltage value, the maximum gate voltage applicable and maximum drain voltage applicable. This chapter content will be a very helpful aid to first realize the modeling outcomes and simulations performed on a 1 μm length long-channel MOSFET with 10 mum gate width and 80 nm gate oxide thickness of drain current as a function of drain voltage with different gate voltages for substrate temperatures $T = 100$ K, 200 K, 300 K, 400 K, and 500 K. It is to be critically noted that direct scaling of these current values at different gate and drain voltage biases at a particular substrate temperatures is not preferable when the channel length after scaling enters from long-channel regime to short-channel regime where various temperature-dependent short-channel effects are well pronounced and proper modeling impact of these effects must be integrated or separately computed to essentially contribute to the simple scaled version of long-channel drive current value of drain current for the n-MOSFET. In this chapter, therefore, we have omitted the additional modeling attributes that emanate from short-channel-induced physical effects and a step-by-step modeling equation-based drain current as a function of drain voltage for different gate voltages are enumerated with modeling outcome plots at different substrate temperatures. Within the long-channel limit of scaling (up to 0.5 μm) the simulation plot outcomes for drain current value can be suitably adjusted and scaled. But for channel length less than < 0.5 μm, different short-channel effects-induced modeling parameter variations make the empirical scaling non-adjustable, threshold voltage being most directly affected and the channel mobility being the other most directly affected because of a shorter channel and consequence of both accentuated vertical gate field and drain field modifying the concept of low drain field with uniform vertical field and uniform channel thickness from source to drain that existed as a precondition for long-channel MOSFET channel mobility extraction. But from a generalized point of view, a constant field-based scaling which has so far been applied to VLSI to ULSI chip integration can be applied to this simulation outcome reported in this chapter considering long-channel drain current values and a suitable window of margin can be defined to account for key short channel effects and a slight process tolerances. The scaled factors must be applied to proportionately define the gate voltage and drain voltage values compatible with device threshold voltage but there is no upper limit of maximum achievable drain current and how much this drain current decreases as a result of scaling induced on channel length. As a result of this projected scaling-induced measurement of drive current, an ITRS road map for maximum achievable drain current for MOSFET device architectures in the vicinity of 10 nm (rough scaling from 1000 nm to 10 nm) can still be targeted with some adjustments of margins of errors by suitable definition of tolerance window as defined above.

3.1.1   MODELING OF SUBSTRATE OR BULK MOBILITY AS A FUNCTION OF SUBSTRATE TEMPERATURE FOR 1 μM CHANNEL LENGTH MOSFET

In order to compute the drain current as a function of drain voltage for different gate voltages of 1 μm channel length n-MOSFET at a particular substrate temperature, channel mobility in the transport direction from source contact to drain contact of a MOSFET is an important modeling parameter. It is correctly found by analysis of scaling evolution of device architectures that the channel mobility if enhanced substantially can more than double the maximum achievable drain current at that particular node. Currently, all the device architectures centric modeling is being carried out at room temperature 300 K. But after all ingenious technology-based architectural modifications of MOSFET including the presently fabricated nanowire FET devices, since the substrate temperature is fixated at room temperature, only marginal enhancement of channel mobility is possible. Whereas same device architecture when tailored to a substrate temperature is close to 100 K, substantial bulk or substrate channel mobility occurs as a result of maximum peak definable phonon scattering-related mobility enhancement at reduced substrate temperature. Although this substrate or bulk mobility does not readily transform into the surface or inversion layer mobility that controls the drain current value being considerably aggravated by the surface roughness scattering, surface-induced carrier confinement, inversion charge screening, and sustained carrier-to-carrier scattering in the case of highly dense inversion carrier density is still a reference substrate mobility value at substrate temperature close to 100 K that is more than two times higher than a corresponding room temperature counterpart can result in a reasonably higher surface or inversion channel mobility value at this reduced substrate temperature and much of the headache of device-engineering professionals to boost the channel mobility value at the most up-to-date device architecture of conventional n-channel MOSFET is peaceably resolved.

In this sub-section of the chapter, first we narrate the modeling equations for calculation of substrate or bulk mobility for electrons in an n-channel MOSFET with different substrate temperatures for a long-channel n-MOSFET [3]. Here the substrate mobility is what essentially the channel mobility with the precondition that the channel is long, the inversion thickness is uniform, the drain voltage is low enough so that drain-induced drain field is negligible and the gate voltage is such that vertical drain field is well below the critical limit that modifies the phonon-induced scattering effect on mobility almost entirely defined by substrate temperature.

$$\mu(T) = \mu_{\min}(T) + \frac{\mu_o(T)}{1 + \left(\frac{N(T)}{N_{ref}(T)}\right)^{\alpha(T)}}, \tag{3.1}$$

where

$$\mu_{\min}(T) = \mu_{\min}(300)\left(\frac{T}{300}\right)^{-0.57} \tag{3.2}$$

$$\mu_o(T) = \mu_o(300)\left(\frac{T}{300}\right)^{-2.33} \tag{3.3}$$

$$\alpha(T) = \alpha(300)\left(\frac{T}{300}\right)^{-0.146} \tag{3.4}$$

$$N_{ref}(T) = N_{ref}(300)\left(\frac{T}{300}\right)^{2.4} \tag{3.5}$$

Since for an n-channel MOSFET the bulk or substrate is p-type, the doping $N(T)$ is constituted by ionized acceptor doping as a function of substrate temperature [2, 3] incorporating incomplete ionization effect at temperatures close to 100 K. The room temperature or 300 K values of N_{ref}, $\mu_{\min}$, μ_o, and α for electron mobility in a p-type bulk substrate are 1.3×10^{17} cm^{-3}, 92 cm^2/V-s, 1268 cm^2/V-s, and 0.91.

$$N(T) = \frac{p_{\mu(T)}}{2}\left(\sqrt{1 + \frac{4N_A}{p_\mu(T)}} - 1\right) \tag{3.6}$$

$$p_\mu(T) = \frac{N_v(T)}{g_A}e^{\frac{E_A - E_V}{kT}}. \tag{3.7}$$

In the above equation, $g_A = 4$, $E_A = 0.045$ eV, and $E_V = 0$ eV taken as reference band energy level for shallow boron doping and the variation of E_A with temperature is omitted due to lack of reference on precisely monitored experimental extraction of acceptor's discrete energy level

E_A as a function of temperature [2].

$$N_V(T) = 2 \left(\frac{2\pi m_p^* k T}{h^2} \right)^{\frac{3}{2}}.$$
(3.8)

Equation (3.6) with the accessories (3.7) and (3.8) are specifically needed to account for incomplete ionization or freeze-out effect of dopant density when the substrate temperature is close to 100–250 K. For substrate temperature greater than 250 K, the following temperature dependent extrinsic doping equation needs to be integrated with Equation (3.6) [2]:

$$N(T) = \frac{N_A}{2} + \sqrt{\left(\frac{N_A}{2} \right)^2 + n_i^2(T)}.$$
(3.9)

In the above equation, $n_i(T)$ is the intrinsic carrier concentration in silicon for different substrate temperatures [2]:

$$n_i(T) = \left(2.510 \times 10^{19} \right) \left(\frac{m_n^* m_p^*}{m_o^2} \right)^{3/4} \left(\frac{T}{300} \right)^{3/2} e^{-\frac{E_G(T) - E_{ex}}{2kT}}$$
(3.10)

$$\frac{m_n^*}{m_o} = 1.028 + \left(6.11 \times 10^{-4} \right) T - (3.09 \times 10^{-7}) T^2$$
(3.11)

$$\frac{m_p^*}{m_o} = 0.610 + \left(7.83 \times 10^{-4} \right) T - (4.46 \times 10^{-7}) T^2$$
(3.12)

$$E_{ex} = 0.0074 \text{ eV}$$
(3.13)

$$E_G(T) = E_G(0) - \frac{\alpha T^2}{\beta + T}.$$
(3.14)

In the above Equation (3.14), the values of $E_G(0)$, α, and β are 1.17 eV, 4.730×10^{-4} eV/K, and 636 K.

3.1.2   MODELING OF DRAIN CURRENT AS A FUNCTION OF DRAIN VOLTAGE FOR DIFFERENT GATE VOLTAGE BIASES AT DIFFERENT SUBSTRATE TEMPERATURES

The threshold voltage V_T as a function of temperature with other device parameters is calculated next [2]:

$$V_T(T) = 2\varphi_B(T) + \phi_{ms}(T) + \frac{t_{ox}}{\varepsilon_{ox}\varepsilon_o} \sqrt{2q\varepsilon_s\varepsilon_o(2\varphi_B(T) N_A(T)}.$$
(3.15)

In Equation (3.15), $\varphi_B(T)$ is the bulk potential as a function of substrate temperature, $\phi_{ms}(T)$ is the metal-to-semiconductor work function difference as a function of substrate temperature, and $N_A(T)$ is the acceptor bulk doping where Equations (3.6) and (3.9) are combined to define bulk doping concentration as a function of substrate temperature from lower than room temperature values to some higher than room temperature values:

$$\varphi_B\,(T) = \frac{kT}{q}\ln\left(\frac{N_A(T)}{n_i(T)}\right) \tag{3.16}$$

$$\phi_{ms}\,(T) = \frac{\phi_M - \phi_s(T)}{q} = \frac{1}{q}\left(\phi_M - \chi - (E_c - E_F(T))\right) \tag{3.17}$$

$$E_c - E_F\,(T) = E_G\,(T) - E_i\,(T) + kT\ln\left(\frac{N_A(T)}{n_i(T)}\right). \tag{3.18}$$

For aluminum metal gate, the value of $\phi_M = 4.28$ eV and $chi = 4.01$ eV for silicon:

$$E_i\,(T) = \frac{E_G(T)}{2} + \frac{3kT}{4}\ln\left(\frac{m_p^*(T)}{m_n^*(T)}\right). \tag{3.19}$$

An enhancement mode n-channel MOSFET drain current can be segmented into three distinct zones, i.e., (i) cut-off, (ii) linear region, and (iii) saturation region. In cut-off, the gate voltage is either zero, negative, or less than threshold voltage V_T and hence there is no current. Although leakage current of very low order of magnitude exists as drain current, for modeling purpose where a textbook drain current as a function of drain voltage is plotted, this current is considered as zero. Essentially, the three regime currents merge at value zero for a drain voltage of 0 V as no forward drift field exists from source to drain that will transport the carriers. In the linear region, the gate voltage is sufficient to turn on the channel and above the threshold voltage V_T and the drain voltage is low enough to be smaller than gate overdrive, i.e., $(V_{gs} - V_T)$. In saturation regime, the drain voltage has considerably increased to be above gate overdrive $(V_{gs} - V_T)$. In this regime even though the drain voltage can be sufficiently higher than gate overdive, the saturation regime drain current is only a function of pinched-off channel voltage at the drain $(V_{gs} - V_T)$ also known as V_{ds} (sat) [2].

For (i): linear region operation of n-MOSFET, $V_{ds} < \left(V_{gs} - V_T(T)\right)$

$$I_D\,(T) = \frac{W}{L}\mu(T)C_{ox}\left[\left(V_{gs} - V_T(T)\right)V_{ds} - \frac{V_{ds}^2}{2}\right]. \tag{3.20}$$

Here, $W =$ channel width (μm), $L =$ channel length (μm), and $C_{ox} =$ oxide capacitance $= \frac{\varepsilon_{ox}\varepsilon_o}{t_{ox}}$ (F/cm^2).

For (ii): saturation region operation of n-MOSFET, $V_{ds} > \left(V_{gs} - V_T(T)\right)$

$$I_D\,(T) = \frac{W}{2L}\mu\,(T)\,C_{ox}\left(V_{gs} - V_T(T)\right)^2. \tag{3.21}$$

The reference long-channel n-MOSFET has a channel length $L = 1\ \mu$m, $W = 10\ \mu$m, $t_{ox} = 80$ nm, and $N_A = 1 \times 10^{16}$ cm^{-3}s.

Using equations combined from (3.1)–(3.20), we arrive at simulation outcomes for drain current as a function of drain voltage for particular gate voltage at different substrate temperatures including 100–500 K showing linear regime of the drain curve. The gate to substrate voltage is $V_{gs} = 2.5$ V. Similar equations combined from (3.1)–(3.21) extend this curve into the saturation region for large enough drain voltages. The plot is shown in Figure 3.1.

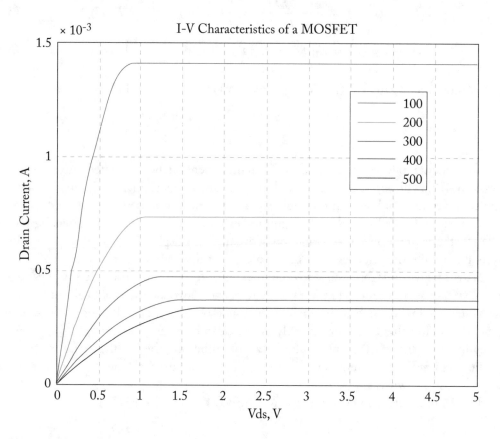

Figure 3.1: Drain current as a function of drain voltage for different substrate temperatures and at a gate bias of 2.5 V. The n-MOSFET has a channel length of 1 μm, width of 10 μm, substrate doping of 1×10^{16}/cm^3, and oxide thickness of 80 nm.

From Figure 3.1, it is clearly evident that as the substrate temperatures of device operation are lowered than room temperature, pronounced increase of mobility impacts the steep rise of the curve during initial turn-on to well-defined linear regime with low enough drain voltage. It is due to this reason that at low enough drain voltage, the channel thickness is uniformly distributed from source to drain and the enhancement of peak channel mobility is then mostly influenced by phonon-scattering-related mobility increase which becomes increasingly conspicuous at lower substrate temperatures than room temperature of operation. In addition, at lower substrate temperature, incomplete ionization leads to fractional ionization of actual bulk substrate doping and has been modeled in the mobility section of this chapter, fractional ionization adds extra contribution to phonon-related mobility peak by reduced Coulomb scattering or ionized impurity scattering. As a result of a sharp nonlinear increase of drain current at lower substrate temperatures visible in this plot, the rather flat saturation region values also show sharp gradation in increments and this is beneficial from the perspective of maximum achievable drain current I_{on} as defined by ITRS. The V_{ds} (sat) value will proportionately increase with an increase of substrate temperatures from 100–500 K. It is due to the reason that for fixed V_{gs}, $(V_{gs} - V_T(T))$ increases as the substrate temperature is increased owing to the fact that at increased substrate temperature, $V_T(T)$ decreases. From a logic switching perspective, we would like to have I_{off} to I_{on} transition (roughly linear to saturation regime inception window) to be defined with an scaled lower V_{ds} range as the supply voltage is scaled. Therefore, we can see from this plot that operations at substrate temperatures close to 100 K will enable this objective of device professionals as for a fixed V_{gs}, $V_T(T)$ can be considerably increased at temperatures in the vicinity of 100 K; as a result, $(V_{gs} - V_T(T))$ can be lower than 1 V and this is the value of V_{ds} (sat) which defines the I_d (sat) or I_{on}. Thus, very low value of V_{ds} (sat) can be designed by modeling of the above equations essentially assisting the voltage scaling but maintaining steep $I_{on} - I_{off}$ transition with additional benefit of increased I_d (sat) or I_{on} at substrate temperatures considerably lower than room temperature operation of conventional n-MOSFET with its scaled projected evolution of architectures.

3.1.3 MODELING OF DRAIN CURRENT AS A FUNCTION OF SUBSTRATE TEMPERATURES FOR DIFFERENT GATE VOLTAGE AND DRAIN BIASES CONDITIONS FOR A LONG-CHANNEL n-MOSFET

In [1] which focused on integration of substrate temperatures with each technology node of architectural evolution, we have enumerated a chapter on how to regulate the die substrate temperature at different logic system subsections through temperature setting and transmitting control unit having access to actuation nodes of different logic sub-circuits defined units assembled in the ULSI chipboard. Detailed CAD simulation can be performed for these logical units to gather the information of drive current, subthreshold current, and threshold voltage requirements for these sub-sections at various stages of operation of the chip including sleep or idle mode operation benefitting static power loss conservation, part of the circuits to be driven slower, part of the circuits to be driven at a higher speed, etc. At different substrate temperatures, values of these CAD-solved drive current, subthreshold current, and threshold voltage must be benchmarked and then, depending on the operation of this chip on a long-term basis, die temperatures for a particular sub-section system of the chip can be suitably engineered to coherently generate all the combinations of the required subthreshold leakage current, drive-on current, and threshold voltage. In this chapter, keeping this objective in mind we have simulated the drain current as a function of substrate temperatures for various gate to source and drain to source voltage biases. The circuits will be mostly under different gate and drain to source voltages of operations generating different sets of current drive values. Spanning these current drive values at different substrate temperatures will allow the device engineer to select the substrate temperature where a specified gate to source and drain to source voltage-defined drain current value will produce maximum advantage for the circuit operation. Figure 3.2 shows this plot. From the plot of Figure 3.2, we can see that for both I_{ds} curves of V_{gs} value of 1.8 V and V_{ds} values of 0.15 V and 0.6 V with linear region of operation, the curves mostly flatter over the temperature range. It is due to the reason that at lower substrate temperatures, the threshold voltage for this device is larger and with a smaller gate over-drive there is less inversion charge density and uniform

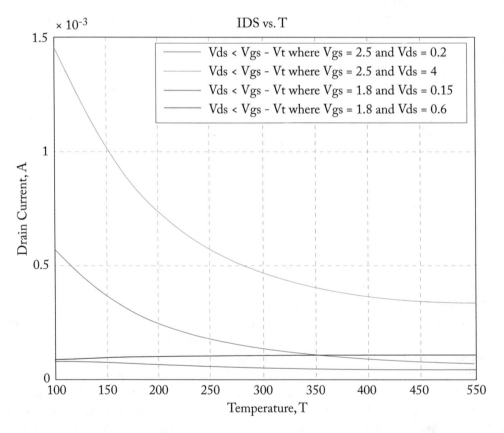

Figure 3.2: Drain current as a function of substrate temperature for different gate to source and drain to source bias voltages. The n-channel MOSFET has the same configuration and critical device parameters, as listed in Figure 3.1.

thickness also aided by lower drain to source voltage. As a result, the mobility is not impacted by either gate to channel vertical field or drain drift field or inversion charge density screening over the whole range of substrate temperatures and as a result the I_{ds} for these two biases of drain and gate voltages, although being lower, is almost non-variant in the entire region of operation of substrate temperatures. These characteristics can be suitably employed for logic circuits operation where gate overdrive is lower due to both voltage scaling and low temperature operation of the circuits and having simultaneously scaled low drain voltage enabling maximum stability of targeted drain current values spanning almost all substrate temperature-operational circumstances. Now let us analyze the I_{ds} curves for V_{gs} value of 2.5 V and V_{ds} value of 4 V showing saturation regime of operation and V_{gs} value of 2.5 V and V_{ds} value of 0.2 V showing linear region of operation. Compared to the I_{ds} curve where V_{gs} value is 1.8 V and $V_{ds} = 0.2$ V, the curve with V_{gs} value of 2.5 V and V_{ds} value of 0.2 V, show marked increase of I_{ds} at lower substrate temperature due to higher inversion charge density due to gate overdrive, but also due these higher inversion charge density, the screening impact on mobility will be also more pronounced and, as a result, the curve shows considerable variation and nonlinearity as a function of substrate temperatures. Therefore, over a targeted tuning range of substrate temperature, stringent control or regulating method has to be built as a circuit so that the required I_{ds} does not deviate too much from reference value at room temperature or 300 K of operation. For the curve where V_{gs} is 2.5 V and $V_{ds} = 4$ V, the I_{ds} value shows the highest point at temperatures closer to 100 K as this curve is now in the saturation region of operation and for a particular gate voltage, saturation region operation of the MOSFET defines the maximum achievable on-current. Although the inversion charge density is same as the curve with $V_{gs} = 2.5$ V, the difference is in the drain voltage, i.e., 4 V instead of 2 V of previous curve. With 4 V, the channel thickness gradually decreases toward the drain and the channel is essentially pinched off for this curve. Because of non-uniform channel thickness, there is a variation of charge density from source to drain and as a result the screening impact on mobility will be more severe for this curve. Therefore, the curve shows maximum variation over the substrate temperatures of operation in spite of the fact that the curve values are maximum at lower substrate temperatures of the MOSFET. From these four curves one final observation that we can deduce is that in order to reap maximum advantage of almost non-variant drain current values over regulated die substrate temperatures, the required gate overdrive or $(V_{gs} - V_T(T))$ value must be low over all temperature ranges and the value of drain voltage also should be sufficiently small and these conditions are ideally the case when voltage scaling for scaled device architectures evolution is considered. For the cases where high enough inversion charge density or higher I_{ds} value is desired or needed for the circuit operation, some control circuit must be built to make sure over a suitable temperature range the variation of drive current is maintained within a tolerable window. Lastly, one point of observation is that curve 3 and 4 both readily decline at high enough substrate temperatures and this is due to sharp reduction of carrier mobility at higher substrate temperatures although gate overdrive, hence inversion charge density as defined by $(V_{gs} - V_T(T))$, is higher for these substrate temperatures of operation. Another attribute of all these four curves are that at lower substrate temperatures incomplete ionization effect results in part of actual substrate doping density unionized and this leads to enhanced mobility and hence higher drive current owing to reduced ionized impurity scattering or Coulomb scattering. At higher substrate temperatures, dopants are fully ionized but intrinsic carrier concentration also rises exponentially with temperatures greater than 400 K leading to slightly increased dopant density than room temperature value and this factor in addition to much enhanced phonon-scattering drastically reduces the channel mobility consequently leading to sharp decline of the drive currents of the upper two curves in Figure 3.2.

3.2 DRAIN CURRENT AS A FUNCTION OF GATE VOLTAGE FOR A FIXED LOW-DRAIN VOLTAGE AT DIFFERENT SUBSTRATE TEMPERATURES FOR THE 1 μM CHANNEL LENGTH n-MOSFET

Although numerous textbooks show the plot of drain current as a function of gate voltage at a fixed low-drain bias for different substrate temperatures, the modeling equations are either simply constructed, empirical or not physically deduced from device surface potential based non-linear transcendental modeling equations. In this book, we have systematically enumerated the surface potential based actual transcendental modeling equations and all the parameters in these equations have their temperature-related evolution precisely extractable from Equations (3.1)–(3.19). When both depletion charge and inversion charge exist in the channel and depletion region in an n-MOSFET at a given gate-to-source voltage, the complete gate-to source voltage V_{gs} and surface band bending potential ψ_s relationship can be expressed as [5]:

$$V_{gs} = V_{FB} + \psi_s + \frac{\sqrt{2\varepsilon_s \varepsilon_o kT N_A}}{C_{ox}} \left[\frac{q\psi_s}{kT} + \left(\frac{n_i}{N_A} \right)^2 e^{\frac{q(\psi_s - V)}{kT}} \right]^{1/2}. \tag{3.22}$$

Here, V_{FB}, N_A, and n_i all are temperature dependent device parameters and V is the channel potential generally fixated at the mid-channel point when the channel thickness is uniform. This condition is fairly established when the drain voltage is very low.

Since the drain to source current is actually determined by the gate-bias and drain-bias modulated inversion charge density, this factor needs to be explicitly separated out from the total charge density underneath the gate of n-MOSFET [5]:

$$Q_i(\psi_s, V) = Q_s - Q_d$$

$$= \left\{ \sqrt{2\varepsilon_s \varepsilon_o kT N_A} \left[\frac{q\psi_s}{kT} + \left(\frac{n_i}{N_A} \right)^2 e^{\frac{q(\psi_s - V)}{kT}} \right]^{1/2} \right\} - \sqrt{2\varepsilon_s \varepsilon_o q N_A \psi_s}. \tag{3.23}$$

Finally [5],

$$I_{DS}\left(V_{gs}, T\right) = \mu\left(T\right) \frac{W}{L} \int_0^{V_{ds}} Q_i\left(\psi_s, V, T\right) dV. \tag{3.24}$$

Combining Equations (3.22)–(3.24), the complete set of curves of I_{ds} as a function of gate voltage for a fixed low-drain bias can be generated at different substrate temperatures. These curves reveal the important and salient attributes of subthreshold leakage current profile and value at low-gate voltage below threshold, the current profile at gate voltage close to threshold voltage and a few thermal voltage above threshold voltage denoting weak inversion and finally at much above gate voltage denoting strong inversion case. The plot is shown in Figure 3.3.

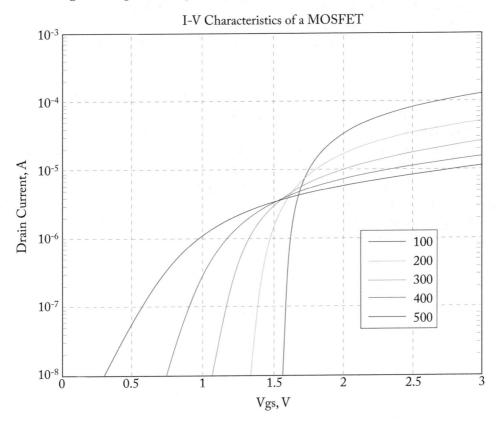

Figure 3.3: Drain current as a function of gate voltage at low enough drain voltage of 30 mV for different substrate temperatures of a 1 μm gate length n-MOSFET. The other device parameters are referred in this chapter from Figure 3.1.

From the above plot, during subthreshold operation we observe that when substrate temperature 100 K is chosen for operation, the steepest fall of current is observed for minute reduction of gate voltage. This is highly desirable from scaled device performance perspectives as for scaled devices, the scaled reduction of gate voltage magnitude will have a very small extension and having substantial reduction of subthreshold current will lower the total leakage current of all the devices integrated in the ULSI chip aiding the overall reduction of chip static power consumption when part of the circuits are idle or minute drain leakage current flow in these circuits. Also from the Figure 3.3, it is evident that for higher substrate temperatures of operation of n-MOSFET, the subthreshold leakage current declines rather gradually and requires much wider gate voltage reduction window below threshold voltage at that particular substrate temperature. For gate voltages higher than threshold voltage, although inversion charge density is smaller for lower substrate temperature operation of n-MOSFET due to decreasing gate overdrive factor $(V_{gs} - V_T(T))$, the much higher substrate mobility (taken as channel mobility at low enough vertical gate field and negligible drain drift field) due to both reduced ionized impurity scattering emanating from incomplete ionization of dopant concentrations and enhanced phonon related scattering will eventually increase the drive current for temperatures of operation closer to 100 K with gate-bias voltage substantially above threshold voltage value. Finally, a point of convergence of all the plots is visible at a gate bias close to 1.5 V and below which the differences in drive current magnitudes rather diminish for all the curves configured at different substrate temperatures. Hence with voltage scaling, the on-current magnitude I_{on} cannot be substantially enhanced due to the lowest room of $(V_{gs} - V_T(T))$ factor and the most improvement will only come from temperature-related mobility enhancement but as these curves reveal, the I_{off} (subthreshold drain current) can be reduced orders of magnitude at reduced substrate temperature of operation and therefore a figure-of-merit I_{on}/I_{off} factor can be enhanced substantially when most device architectures at scaled node are tailored to operate at a substrate temperature considerably below room temperature of operation.

CHAPTER 4

Simulation Results Substrate Mobility and On-Channel Mobility of Conventional Long-Channel n-MOSFET at Substrate Temperatures 300 K and Below

In Chapter 3, we showed simulation results of on-state drain current and subthreshold drain current as a function of substrate temperatures and also highlighted the utmost importance of operating today's ULSI circuits at substrate temperatures considerably below 300 K to rip the benefit of reduced subthreshold drain current impacting substhreshold and off-state leakage current minimization and simultaneously moderately augmenting the on-state drive current at reduced substrate temperatures where voltage scaling allows for the reduction of the voltage on the gate and drain to the extent that $(V_{gs} - V_T)$ the gate overdrive is not too overly minute and there is sufficient drift field allowed by not too aggressive scaling on drain voltage V_{ds}. In this chapter, we focus on similar simulation outcomes at different substrate temperatures on (i) bulk substrate mobility of electrons in p-substrate and (ii) electron inversion layer mobility when an n-channel MOSFET is fabricated on p-type substrate. In today's scaled MOSFET, it is the inversion layer mobility that suffers horrendously to the extent that a non-scalable plateau has emerged on maximum achievable inversion layer mobility limiting the value to about $100 \, \mathrm{cm^2/V}$-s. At room temperature or 300 K, such a small value of inversion layer mobility will not keep the drive current in commensurate with device scaling and the ULSI clock speed for gazillions-of-instruction-per second of microprocessor CPU will definitely need to be revised and trimmed from the highly expected ITRS road map. Operation of today's most state-of-the-art device architecture such as nanowire MOSFET at reduced substrate temperatures close to 100 K will first enhance the phonon-scattering limited substrate mobility for these nanowire substrates at these reduced than room temperature temperatures and concurrently the fraction that goes to determine surface or inversion layer mobility will also be higher at these reduced substrate temperatures for these n-channel nanowire MOSFETs. With a view of this wide range of intrinsic device operational benefit on substrate mobility that can be configured by ingenious engineering of substrate temperatures, we first show simulation outcomes on substrate electron mobility on p-type silicon substrate as function of substrate acceptor doping concentrations at different substrate operational temperatures.

4.1   ELECTRON MOBILITY IN p-TYPE SUBSTRATE OF SILICON VARYING WITH SUBSTRATE ACCEPTOR DOPING CONCENTRATIONS FOR DIFFERENT SUBSTRATE TEMPERATURES

The fundamental modeling equation for electron mobility in p-type substrate with different acceptor doping concentrations is the (3.1) as detailed in Section 3.1. Combining (3.2)–(3.14), the temperature dispersion or evolution relationships can be established for all the parameters of Equation (3.1). Equation (3.1) when plotted for electron mobility μ as a function of acceptor doping concentrations (N_A) for different lattice or substrate temperatures result in the following simulation outcome plot.

From Figure 4.1, it is clearly evident that at low near intrinsic doping concentrations, the variation of mobility with doping is almost nonexistent and the mobility, mostly defined by phonon scattering, reaches substantial engendered magnitudes at substrate temperatures close to 100 K Due to incorporation of incomplete ionization effect dominant at reduced substrate temperatures, the reduced screening from ionized acceptor doping concentrations add up factor resulting from reduced Coulomb scattering enhanced mobility to the natural phonon-scattering related mobility enhancement at these reduced substrate temperatures. As the acceptor doping concentrations exceed a few $10^{16}/cm^3$, the electron mobility decreases rapidly for all substrate operational temperatures and at regions of substrate doping above $10^{18}/cm^3$, the benefit on mobility that comes out of reduced substrate temperature operation ceases to exist with the mobility virtually becoming a constant very low value of the order 200 cm^2/V-s even when reduced substrate temperature operation is considered. Due to increased Coulomb scattering resulting from high concentrations of ionized acceptor dopants, all the advantages of phonon-scattering enhanced mobility at reduced substrate temperatures gradually disappear. Moreover, the incomplete ionization effect has lesser impact with progressive reduction of substrate temperature as the device doping is increased or higher and origination of Coulomb scattering-related mobility fall-off or reduction starts early in the device doping range for reduced substrate temperatures of operation. In this connection we note that if conventional long-channel n-MOSFET had to be scaled in a natural way up to 10 nm technology node, the requirement of source and drain doping region isolation or separation would necessitate a highly aggressive substrate doping of the orders of few $10^{18}/cm^3$ with the natural outcome of a very low substrate mobility converted

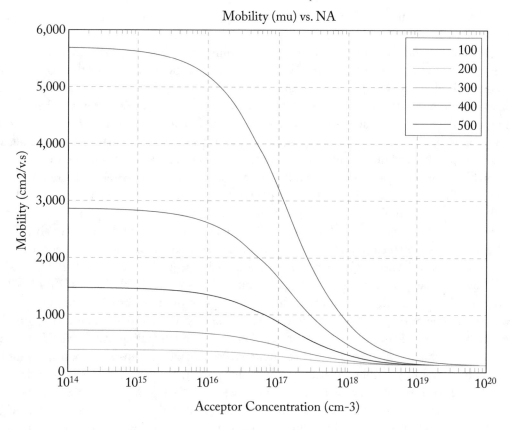

Figure 4.1: Electron mobility as a function of *p*-type substrate acceptor doping concentrations for different substrate temperatures.

inversion layer mobility which cannot be scaled up any further by reduced substrate temperature operation. Fortunately, device engineers and practicing professionals were cognizant of these facts and architectures of double-gate, tri-gate, Fin-FET, and nanowire devices were developed which operate on the principle of near intrinsic device doping close to a few $10^{14}/cm^3$. From the plot of Figure 3.1, we see that mobility values are significantly higher at these intrinsic doping for all reduced substrate temperatures than 300 K and additionally due to negligible Coulomb or ionized impurity scattering, these mobility values once set by substrate temperature of choice will not vary in a moderate amount if substrate doping variation is considered such as non-uniform doping from source to drain junction MOSFET or raised pocket or halo based doping at source and drain regions or even highly beneficial retrograde doping with lower concentration at the surface and gradually engineered higher concentration toward the substrate of an *n*-channel MOSFET. The closest resemblance that we find to Figure 4.1 through literature reference is the highly effective empirical modeling based electron mobility computed and extracted in reference to Baccarani's et al. paper on [91]. This reference article shows the extracted simulation outcome of minority electron mobility in p-type substrate in Figure 9 of this article [91] that is a close replica of the Figure 4.1 generated by the author of this book. Additionally, their plot required use of many empirical parameters with their temperature evolution as listed in Table II of this reference article. Compared to the modeling equations and empirical fitting parameters that the authors of the reference article have employed, the equations that the author of this book have utilized are completely configured based on material specific device physics for p-substrate silicon and the parameteric temperature based dispersion or evolution relationship have been configured from device electrostatics of silicon material. Therefore, while Figure 9 of Baccarani et al. with its Table II [91] of the article can be efficiently integrated in a CAD-based device simulation modeling software, the plot of Figure 4.1 of this book with the modeling equations outscored in this book will serve as an essential tool providing alternative assistance of choosing the mobility value at a particular substrate doping at the respective device operational substrate temperature with analytical modeling equations computable in a considerably shorter time span from industry standard numerical software tools such as MATLAB or MathCAD. To finish this section, the author would like to draw the attention of device engineers about some of the aspects of substrate or bulk mobility when high substrate doping in the degenerate limit needs to be considered in relation to continued scaling of MOSFET devices. For the circumstances of choices of device architectures when high substrate doping cannot be avoided, bulk mobility is not only affected adversely by many body scattering related reduction of mean free path of traveling electron carriers between scattering but a possible increment of effective mass of electron can result due to many body induced warping of conduction band edge minima which has already the feature of non-parabolicity. A possible increment of effective mass of electron when degenerate bulk doping is the case will additionally reduce mobility apart from many body-electron scattering interactions. The drastic reduction of bulk mobility in presence of heavy substrate doping eventually will transform the maximum achievable surface or inversion layer mobility below 100 cm^2/V-s because of effects like quantum confinement and extreme surface inversion carriers scattering. Second, nowadays MOSFET built on wafer is not only aligned in (100) directional wafers but also in (110) and (001) directional wafers where 2-D energy confinement splits the conduction band valleys and alter the longitudinal and transverse effective masses of electrons that eventually determine the 3-D effective masses for electrons as carriers in substrates with different crystallographic directions. Therefore simulation outcomes like Figure 4.1 should be generated for p-substrates oriented in (110) and (111) directions and proper modeling equations aided by dispersion relationships with substrate temperatures must be established for electron effective masses in these directions of substrate orientations taking guidelines from experimental derivations or measurements. Equations (3.11) and (3.12) which are standard textbook forms cited for (100) oriented p-substrate (intrinsic electron and hole carriers) need to be remodeled if we would like to reap benefit from substrate mobility-converted inversion layer mobility improvements with wafer alignments in other directions.

4.2   SIMULATION RESULTS OF ELECTRON CARRIER MOBILITY AT THE SURFACE OF AN n-CHANNEL MOSFET FOR DIFFERENT SUBSTRATE TEMPERATURES

One of the much needed bonanza of device carrier transport at lower substrate or lattice temperatures is the enhanced carrier mobility extracted through carrier transport characteristics with near intrinsic substrate doping when the device is operated at lower temperature. At temperatures of the order of 100 K or below, the multiple factor of enhancement of carrier mobility can be as large as 4–5 times when the mobility engenderment factor is governed by phonon scattering limited doping concentration and away from the onset of Coulombic scattering interaction zone. Unfortunately, very different extraction results when electron transport near the interface (surface channel mobility) of an n-channel MOSFET with SiO_2 gate dielectric is considered. Due to pronounced surface scattering related mobility peak reduction from its phonon scattering limited maximum peak or plateau, the actual carrier transport related channel mobility of experimentally fabricated devices concentrated upon 100–500 cm^2/V-s even when reduced substrate temperature operation condition is adopted by device engineering technology application professionals. Being cognizant of the fact that surface channel mobility can be significantly attenuated from its phonon scattering limited bulk or substrate mobility value owing to the condition of interface layer scattering and also due to the progressive alterations of mean free path between scattering of electrons with themselves and underlying depletion region ions as a result of thin inversion layer and underlying depletion region both being function of gate-to-channel electric field and drain-to-source drift electric field which gives rise to additional surface field related mobility reduction, device engineers have concentrated on possible technology modifications starting from 90 nm node to boost channel carrier mobility like introducing compressive and tensile strain materials in the channel, buried or embedded stressors at the source and drain and channel regions, introducing high-k metal gate (HKMG) technology, capping the high-k gate with stressors, etc., along with different orientations of substrate and channel transport directions. But after successful implementation of all these technologies the channel or surface interaction-dominated electron mobility for an n-channel MOSFET could not be raised more than 200 cm^2/V-s at room temperature in the vicinity of technology node of 22 nm or below. It is because the surface field-induced attenuation factor that is responsible for mobility reduction in experimentally fabricated devices is empirically related to mobility modeling equations and cannot be properly linked to its actual device physical attribute since surface scattering of transport is really a time-dependent stochastic event and is also dependent on the near atomically smoothened silicon-SiO_2 interface which with today's fabrication instruments is a giant norm to achieve or execute in production factory. Hence, all the above technological modifications gave way to only marginal improvement of surface channel mobility of electrons in an n-channel MOSFET and alternate device technology was in the asking since we entered the technology node 18 nm or below when the channel mobility of production quality n-channel MOSFET die has even gone down below 100 cm²/V-s as reported in many IEDM papers. The author of this book believes and does simulation outcomes that show that if the bulk or substrate electron mobility can be enhanced 4–5 times (of the order of 6000 cm²/V-s for Figure 4.1 at $T = 100$ K) by operating the device at a much reduced operating temperature taking room temperature operation as a basis, even after surface field-related attenuation empirically modeled at these reduced substrate operating temperatures, the phonon-scattering related mobility peak will not be trimmed to the extent that it will fall to 100–200 cm²/V-s rather the values will be in the range 500–600 cm²/V-s considering that intrinsic doping is chosen and oxide thickness and applied gate and drain voltages are such that the surface channel related mobility defined by peak vertical field is made to exist close to phonon-scattering related zone and not inclined to interface roughness scattering zone (accentuated by high bulk doping and extremely thin gate oxide) and Coulomb scattering-related zone (accentuated by too low doping and too low voltage on the gate indicative of near threshold or subthreshold region of operation). It is thus legitimate to expect that with consideration of short-channel scaling factors and other factors like quantum two-dimensional confinement and boundary layer scattering existing in nanowire n-channel high-k metal gate MOSFET in addition to surface field-related mobility attenuation factor, the near 6000 cm²/V-s electron mobility as shown in Figure 4.1 at substrate temperature 100 K in the bulk of a p-silicon substrate will at least generate a mobility of the order of 500–600 cm²/V-s with batch fabricated n-channel MOSFET die in a wafer with production standard oxide-silicon interface. The device engineers are also aided by the fact that the technology node enters 10 nm node or below that the substrate or bulk doping of these nanowire devices are near intrinsic level allowing volume inversion (less Coulombic scattering of inversion layer electrons by dopant atoms in the depletion region) and imparting more amenability to reduced substrate temperature operation-based mobility enhancement with more controlled surface field related mobility reduction (near absence of interface roughness scattering and increase of mean-free path between scattering due to less probabilistic carrier-to-carrier and carrier-to-depletion-region-ions scattering processes, a factor governed by channel doping condition and operating substrate temperature). Before we narrate the simulation outcomes of inversion layer mobility for different substrate temperatures as a function of effective vertical electric field, some insightful remarks are in the offing in relation to the discussion we have presented in this sub-section. The Drude's mobility model has the representative equation $= \frac{q\tau}{m^*}$. In order to increase inversion layer mobility, one option is to increase the mean-free path between scattering or increase τ and the other option is to decrease the effective mass of carrier or m^*. From surface mobility standpoint, the abrupt interface at the Si-SiO₂ layer causes disruption in the periodic arrangement of silicon atoms at the interface calling for remodeling of the E

(energy)-k (wavevector) diagram to extract the effective mass of electron from the conduction band minima curvature which obviously has a much strained and constrained shape compared to the bulk level E-k diagram from where we usually report the effective mass values. The textbook definitions have thus far called these effective masses different from their bulk neighborhood as conductivity effective mass. Both electrons and holes have less conductivity effective masses compared to density gradient effective mass values. This will tend to up-shift the mobility value of the Drude's equation referred to here. Unfortunately, scattering at the surface makes the electron carrier transport with loss of directional momentum converted drift velocity and the τ value is significantly decreased compared to relative decrease of m^*. Therefore, overall mobility value degrades. Although lower-temperature operation increases effective mass slightly, employing the effective mass-temperature dispersion relationship quoted earlier for silicon, it significantly increases the mean-free path between scattering resulting in longer τ and as a result even the inversion layer surface mobility can be increased to a desired proportion by suitable adjustment of lower temperature operation of today's scaled device architecture. The other increasingly desirable need where mobility can be enhanced is the ordered doping of atoms resulting in ordered arrangement of dopant atoms in the depletion region. Using scanning tunneling microscopy along with controlled dose of ion-implantation, a near ordered array of dopant atoms can be placed close to the inversion layer in the channel region of the MOSFET constituting the depletion region. Now since the scattering pattern with dopant atoms of the transporting electrons also undergoes systematic and moderately spaced sequence, the overall weighted average on τ gives rise to much larger scattering time even at room temperature. This technological integration will additionally increase the scattering time at lower-operational substrate temperature owing to the fact that some of the ordered dopant atoms are unionized which reduces the total number of ionized atoms in the depletion region and therefore induces less ionized impurity related Coulomb scattering with slight to moderate increase in mobility. The author here wants to make a point that instead of devising the technology for reduction of effective mass, a more concerted goal should be directed toward inventing the technology to increase the mean-free path between scattering or longer scattering time noting that proportional increase factor of τ is much more definable than decrease factor of effective mass. Now we systematically document the modeling equations developed and simulation outcomes of the surface channel mobility of electrons in an n-channel device with acceptor doping density of $10^{16}/\text{cm}^3$ for two different substrate temperatures 300 K and 200 K, as a function of effective gate driven vertical field.

4.2.1 MODELING EQUATIONS FOR EXTRACTION OF SURFACE MOBILITY AS A FUNCTION OF VERTICAL EFFECTIVE FIELD

The inversion channel mobility at low drift field (low-drain voltage) is commonly calculated from the following equation [5]:

$$\mu_{eff} = \frac{L}{W} \frac{g_d(V_g)}{Q_{inv}(V_g)}. \tag{4.1}$$

In Equation (4.1) μ_{eff} is the effective channel mobility, L is the channel length of n-MOSFET, W is the channel width of n-MOSFET, and $g_d(V_g)$ is the conductance defined by derivative of drain current with respect to drain voltage at the linear region of I_d (drain current)–V_{ds} (drain voltage) plot for different gate voltage biases. $Q_{inv}(V_g)$ is the sheet inversion charge density in the channel of an n-MOSFET also dependent on gate bias at low enough drain bias signifying the charge density is spatially uniform in the channel region from source to drain. Equation (4.1) gives the effective mobility value as a function of gate voltage which can be further converted to effective vertical field.

In order to calculate $g_d(V_g)$, first from the bulk charge theory as laid out by J. R. Brews, we first use the following current equation for drain current as a function of drain voltage for a set of gate voltages and generate a complete family of curves extending from linear region to saturation region operation of n-channel enhancement mode MOSFET [5]:

$$I_{ds} = \mu_{eff} C_{ox} \frac{W}{L} \left[\left\{ \left(V_g - V_{fb} - 2\varphi_B - \frac{V_{ds}}{2} \right) V_{ds} \right\} - \frac{2\sqrt{2\varepsilon_{si}\varepsilon_0 q N_A}}{3C_{ox}} \left\{ (2\varphi_B + V_{ds})^{3/2} - (2\varphi_B)^{3/2} \right\} \right]. \tag{4.2}$$

The device engineers are familiar with the above equation from their post-graduate study of MOS capacitors and MOSFETs. All the terms have been explained before and bear the exact connotations as described previously. For a specified acceptor doping N_A (/cm^3) and oxide thickness t_{ox} (nm), I_{ds} in the above equation can be expanded as a function of very low to moderate to high drain voltage selecting a particular gate voltage. The equation in itself contains the required approximated closed form expression for linear region ($V_{ds} < V_{gs} - V_T$) and saturation region ($V_{ds} > V_{gs} - V_T$) of the n-MOSFET where V_T equation has been described in detail in [1]. All the relevant parameters that have dependence on substrate temperatures in the above equation when modeled give rise to a family of curves of drain current-drain voltage with respect to different gate voltages at a particular substrate or lattice temperature. We show a representative simulation outcome of drain current as a function of drain voltage from linear region to saturation region operation of n-MOSFET for high enough gate voltage when the entire device is operated at different substrate temperatures from 100–500 K.

From Figure 4.2, we can identify the near linear rise at each of the five temperature settings when drain voltage is sufficiently low and hence the slope taken in the vicinity of this linear region will give the value of $g_d(V_g)$ for each of these temperatures. If V_{ds} is low enough like 10–30 mV, the variation of $g_d(V_g)$ locally will be almost indistinguishable.

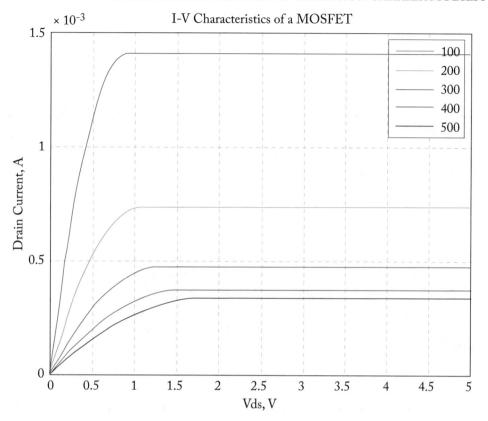

Figure 4.2: Drain current as a function of drain voltage for different substrate or lattice temperatures of operation of the device at a high enough gate voltage.

In order to calculate $Q_{inv}(V_g)$ in Equation (4.1), we first isolate it from the total surface charge by subtracting the depletion charge expression [5]:

$$Q_s\left(\psi_s\right) = \frac{\sqrt{2\varepsilon_{si}\varepsilon_0 kTN_A}}{C_{ox}} \left[\frac{q\psi_s}{kT} + \frac{n_i^2}{N_A^2}e^{\frac{q\left(\psi_s - V\right)}{kT}}\right]^{1/2} \tag{4.3}$$

$$Q_d\left(\psi_s\right) = \sqrt{2q\varepsilon_{si}\varepsilon_0 N_A\left(\psi_s + V\right)} \tag{4.4}$$

$$Q_{inv}\left(\psi_s\right) = Q_s\left(\psi_s\right) - Q_d\left(\psi_s\right). \tag{4.5}$$

The values of ψ_s for different V_g values are related by [5]:

$$V_g = V_{fb} + \psi_s + \frac{\sqrt{2\varepsilon_{si}\varepsilon_0 kT N_A}}{C_{ox}} \left[\frac{q\psi_s}{kT} + \frac{n_i^2}{N_A^2} e^{\frac{q(\psi_s - V)}{kT}} \right]^{1/2}. \tag{4.6}$$

The potential V is the average channel potential which for low enough drain voltage turns out to be $V_{ds}/2$. Equations (4.3)–(4.6) relate Q_{inv} for a particular gate voltage required in Equation (4.1). Finally, for a specific gate voltage its corresponding surface potential value ψ_s establishes the effective vertical gate-driven electric field at low enough drain voltage by the following equation [5]:

$$E_{eff} = \frac{\left(Q_d(\psi_s) + \frac{Q_{inv}(\psi_s)}{2} \right)}{\varepsilon_{si}\varepsilon_0}. \tag{4.7}$$

For different substrate or lattice temperatures, the temperature-dependent parameters of the equation set (4.3)–(4.7) are known beforehand and their dispersion evolution are also modeled appropriately, some of which have been mentioned previously as well as in [1]. At room temperature or 300 K, the well-known Takagi universal effective surface mobility model is given below which is used for extracting the plot of effective surface mobility as a function of effective vertical field for substrate doping $N_A = 1 \times 10^{16}/\text{cm}^3$:

$$\mu_{eff} = 32500 \times E_{eff}^{-0.33}. \tag{4.8}$$

Although the author initially opted for all the surface mobility plots as a function of effective vertical field for substrate temperatures 100 K, 200 K, 300 K, 400 K, and 500 K, due to time constraints and limited research support, the room temperature surface mobility plots using Equations (4.1)–(4.7) and at temperature of 200 K are illustrated here to attract the reader's attention to the difference of peak mobility that can be engineered at a low enough substrate temperature by judiciously selecting the required effective vertical field. For any temperature T different than room temperature 300 K, the modeled equation is developed by recognizing that at low enough vertical field the near peak of effective mobility is related to room temperature mobility by the lattice impurity scattering-related temperature-dependence exponent, i.e.,

$$\mu_{eff}(T) = \mu_{eff}(300) \left(\frac{T}{300} \right)^{-1.5}. \tag{4.9}$$

In the above equation, $\mu_{eff}(300)$ is the extracted value of (4.8) with very low vertical gate field. Now at $T = 200$ K, the low field surface mobility value is essentially the previously generated bulk mobility value for the referenced doping of $10^{16}/\text{cm}^3$. Therefore, a peak adjustment setting factor is introduced as

$$C_1 = \frac{\mu_{eff}(300)}{\mu_{bulk \to eff}(200, E_{eff})}. \tag{4.10}$$

After inclusion of adjustment factor C_1, the final expression of effective surface mobility at a substrate temperature T other than 300 K is given by:

$$\mu_{eff}(T, \varepsilon, \eta) = C_1 \times \mu_{bulk \to eff}(T) \times \left(\frac{T}{300}\right)^{-1.5} \times \varepsilon^{-\eta}. \tag{4.11}$$

Here, the surface field-defining exponent η has the following temperature dependent modeling expression:

$$\eta(T) = a\left(\frac{300}{T}\right) + b\left(\frac{300}{T}\right)^2. \tag{4.12}$$

In order to extract the values of a and b, we utilized the highly referenced Takagi universal mobility plot at two different substrate temperatures: $T = 300$ K and $T = 77$ K The asymptotic values of η at $T = 77$ K is 2 and at $T = 300$ K is 0.33. From Equation (4.12), the value of η at substrate temperature 200 K used for simulation outcome is computed as 0.54. The proper values of η at different substrate temperatures of operation of MOSFET should not be deduced empirically rather from experimental characterization of many dies over a wide range of substrate doping distributions just like the Takagi plot demonstrated from experimental validation of two η values at $T = 77$ K and $T = 300$ K. The modeling equation of (4.12) will not produce exact surface mobility trend with vertical field as detailed experimental characterization at $T = 200$ K is missing to validate the modeled simulation plot but the nonlinear relationship chosen for Equation (4.12) essentially reinforces the fact that due to added surface scattering at sufficiently lower temperature than 300 K, the surface field reduction exponent η will be intimately dependent on these transport characteristics giving rise to a non-proportionate reduction factor.

Now we systematically discuss how the effective mobility vs. electric field relationships for different substrate doping at two substrate temperatures (300 K and 77 K) as developed by Professor Dr. Shin-ichi-Takagi and his collaborators at the University of Tokyo can be fairly regenerated by incorporating the above Equations (4.1)–(4.12), as developed by the current author. Additionally, Equations (4.1)–(4.12) can be parametrically engineered to simulate the effective vertical field as a function of effective vertical gate field for temperatures close to 300 K, i.e., 200 K, 150 K, 100 K, etc., consequent to the fact that I have elaborated before that low-temperature operation of MOSFETs does not necessarily mean cryogenic temperature operation of considerably low-temperature operation of a MOSFET device which will require expensive accessorial cryo-electronics or cryo-coolers whereas if the substrate temperature is close to 300 K, die-attached solid-state micro-coolers with low power drive should be sufficient to enable the MOSFET operation for better drive efficiency and from the perspective of leakage power consumption, as described in [1]. At a given V_g, for a particular temperature $T = 300$ K and substrate doping $N_A 10^{16}$/cm^3 with oxide thickness $t_{ox} = 80$ nm, Equation (4.6) is used to extract ψ_s. Then combining Equations (4.3)–(4.5) with the value of ψ_s extracted from Equation (4.6), we compute the inversion charge density Q_{inv}. Now at very low electric field at $T = 300$ K or at very low V_{gs} value, the mobility value is essentially determined by phonon-scattering related peak value which is essentially the bulk substrate mobility value as the surface field is very low to impact this value chosen as peak surface mobility. After E_{eff} is calculated from Equation (4.7) for a particular ψ_s, the first data sample of effective mobility μ_{eff} (i) with $i = 1$ at low enough gate field is calculated from Equation (4.8). Now we use Equation (4.2) to generate I_{ds} as a function of V_{ds} with the low-gate field determining V_{gs} and take a slope of the curve at very low $V_{ds} = 30$ mV denoting a linear region of operation. The slope of this curve is the g_{ds} value. For a new gate field as determined by new V_{gs}, first we adjust g_{ds} from the curve of $I_{ds} - V_{ds}$ for $i = 1$. Then we use Equation (4.1) to generate a new μ_{eff} ($i = 2$) which will not be accurate as g_{ds} value associated with it has been derived from a previous point $I_{ds} - V_{ds}$ curve. No g_{ds} value is recomputed on $i = 2$ $I_{ds} - V_{gs}$ curve from Equation (4.2) and new μ_{eff} ($i = 2$) is calculated from Equation (4.1). In another iteration, this new μ_{eff} ($i = 2$) is used to generate new g_{ds} as per Equation (4.2). Another calculation of μ_{eff} ($i = 2$) is done from Equation (4.1). The difference of these two values (second iteration and third iteration) of μ_{eff} ($i = 2$) is noted and the derivative point on $I_{ds} - V_{ds}$ curve for the new V_g (sample point 2, $i = 2$) is adjusted until the difference of the two μ_{eff} ($i = 2$) values is fairly negligible. At this point, we have found the μ_{eff} ($i = 2$) value at a new gate field. For μ_{eff} ($i = 3$, data sample #3) onward, we repeat this procedure up until the gate field is fairly large that the mobility diminishing asymptote essentially replicates the -0.33 factor as given in Equation (4.8) where the gate overdrive is sufficient multiple of threshold voltage V_t with low enough drift field or low drain voltage. The simulated outcome of effective carrier mobility at the inversion channel surface of an n-MOSFET at temperature 300 K for substrate doping $N_A = 10^{16}/cm^3$ as a function of effective vertical gate field for sufficiently low drain field is shown in Figure 4.3.

As we can see from Figure 4.3, at low enough gate field or near threshold region, the peak of the mobility is close to 900 cm²/V-s which is close to bulk substrate mobility value at this substrate doping of $10^{16}/cm^3$. As the gate field is increased, the asymptotic decrease as per -0.33 exponent for $T = 300$ K is evident from this figure with the mobility value approaching 100 cm²/V-s at a high enough gate field with sizeable surface roughness scattering. These values or trends are fairly close to Takagi's universal mobility plot but we have used the device physically configurable n-MOSFET equations to extract the surface mobility rather than using empirical fitting parameters. However, one advantage Dr. Takagi's group had was that they validated their modeling equation's derived mobility values with experimental characterization.

For another temperature such as $T = 200$ K used for simulation of effective channel electron mobility as a function of gate effective field, Equations (4.11) and (4.12) have to be used at a fairly low electric field to define the first sample point for μ_{eff}. Then a temperature-related value update of all the parameters of Equations (4.2)–(4.7) has to be done. Using the same sequential computing steps as enunciated for $T = 300$ K operation, the channel mobility values can be calculated for substrate temperature $T = 200$ K as a function of gate effective field. We show this plot in Figure 4.4.

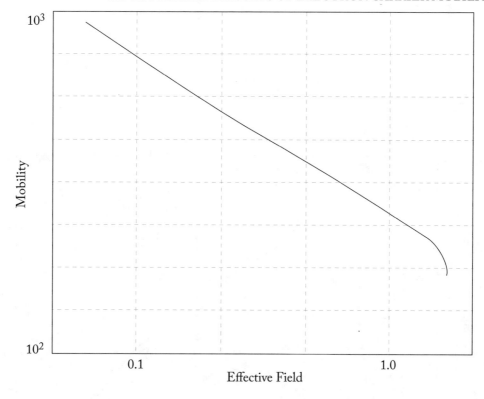

Figure 4.3: Effective electron carrier mobility as a function of an effective field. (The units of mobility are cm²/V-s and effective vertical field are V/cm. The substrate temperature is 300 K.)

From Figure 4.4, the effective channel mobility peak has been much enhanced at substrate temperature $T = 200$ K at low enough gate field where threshold region of operation of n-MOSFET is located. At this region, the mobility is aided by both reduced phonon-scattering and reduced Coulomb scattering from incomplete ionization of dopants in the depletion region underneath the channel inversion layer. Also, the mobility decreases at a faster asymptotic rate compared to $T = 300$ K operation as device physics of transport necessitates this decline due to increased surface roughness scattering setting in at a comparatively lower gate field of operation compared to $T = 300$ K gate field dependent surface roughness scattering initiation. Figure 4.4 is a new illustration of a plot compared to the plot provided by Takagi et al. for $T = 77$ K. The only qualifying aspect of Figure 4.4 is its need to be validated by experimental fabrication of this n-MOSFET and characterizing the effective mobility values with subsequent fitting with this modeled curve of Figure 4.4.

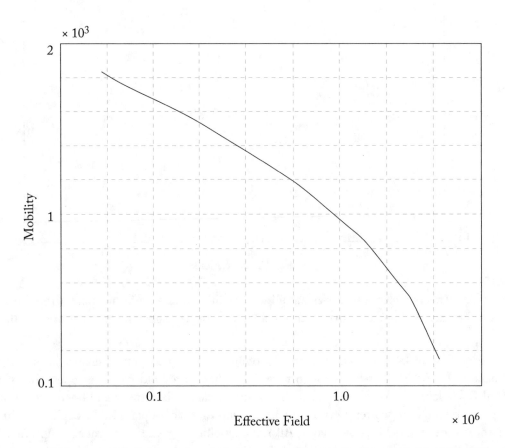

Figure 4.4: Electron channel mobility as a function of vertical effective field. (The substrate temperature is 200 K and mobility unit and effective field units are same as in Figure 4.3.)

CHAPTER 5

Simulation Outcomes of Subthreshold Slope Factor or Coefficient for Different Substrate Temperatures at the Vicinity of a Subthreshold Region to Deep Subthreshold Region of a Long-Channel n-MOSFET

One of the very important performance benchmark-driven device parameters is the inverse subthreshold slope (mV/decade) of an n-channel MOSFET which determines the steep switching characteristics of the device during fast turn-on and turn-off and hence contributes to the energy dissipation or transition factor when logic devices such as ring oscillators, inverters, NAND gates, NOR gates, Multiplexers, Domino-n logic gates, etc., which are subject to femtosecond time period clock gating which requires extremely low-subthreshold slope factor or the fastest switching characteristics from digital performance standpoint. At room temperature or 300 K, the standard subthreshold slope comes out to be 60 mV/decade or $S = 2.3nkT/q$, where S is inverse subthreshold slope, kT/q is the thermal voltage, and n is the subthreshold slope factor. At room temperature, n is equal to 1. That is an ideal value of n when the depletion width is almost non existent or depletion capacitance C_D is zero so that charging and discharging of C_D during switching contributes no additional term to n value and the point of transition from turn-on to turn-off is entirely driven by thermionic energy-based conduction electron recovery at the small gate-to source side energy barrier when the turn-off mode is encountered. Even though at $T = 300$ K ideal subthreshold slope is 60 mV/decade, it posed significant challenge to device professionals to come close to this benchmark starting from 1 μm long-channel device to today's extremely scaled device when the channel doping needed to be raised at every node of scaling to enable efficient device performance and alleviate detrimental short-channel performance features like source-to-drain punchthrough leakage current, drain-induced barrier lowering, and drain field extension to the source side. As a result of this requirement of enhancement of channel doping values, the depletion region width becomes narrower giving rise to increased C_D and an increase of C_D/C_{ox} subsequently enhancing the value of n above than 1. Although C_{ox} increased for every technology node considering scaled reduction of oxide thickness or t_{ox}, since C_D becomes comparable to C_{ox} due to its precipitated reduced depletion width at technology node scaling evolution, the reduction of n in line to 60 mV/decade was kind of a technology mirage or illusion for device professionals as up until them devices like double-gate MOSFET, FDSOI MOSFET, Fin-FET, Tunnel-FET, and gate-all-around nanowire FET architectures were developed which enabled the attainment of near 60 mV/decade subthreshold slope or even lower than this value at room temperature by dint of gate integrity control, near intrinsic channel doping and making the subthreshold operation tunneling based in the case of Tunnel FETs rather than thermionic emission based for other device architectures cited in this connection. When a device is operated at lower substrate temperature, it forcefully impacts the subthreshold slope values in two conspicuous ways. First, the thermal voltage term kT/q is directly reduced with lower substrate temperature operation as T is reduced. Second, the subthreshold slope factor n contains the term C_D which has a dependence of depletion layer width at subthreshold region operation of MOSFET. This depletion width has a direct square root dependence on surface band bending potential ψ_s and an inverse square root dependence on doping value. As the temperature of operation is reduced, incomplete ionization sets in curbing the amount of dopant ionizations to free electron carriers reducing the value of ionized dopants. Moreover, as the temperature of operation is reduced, the surface band bending potential at the reduced gate voltage defining subthreshold region of operation is simultaneously increased and both these device intrinsic properties first increase the depletion width ideally reduce the value of n by attenuating the value of C_D and hence subthreshold slope S. With low to moderate doping which will be required for today's ultra-scaled MOSFET devices to control drain field penetration beneath the channel to the source while gate-to-body integrity is preserved, the eventual S value coming out of our detailed analytical surface potential based extraction is around 32 mV/decade when the device is operated at 100 K which is highly beneficial for devices that are meant to operate with extremely low-supply voltage operation or operational scenarios present in ASIC based microcontroller applications where subthreshold logic technologies are employed.

For a suitably selected V_g or gate voltage defining that V_g is sufficiently lower than threshold voltage of our n-channel MOSFET whose device parameters defining threshold voltage at room temperature 300 K have been stated previously, the surface band bending potential ψ_s is determined from Equation (4.6) in the presence of all temperature-dependent modeled parameters that have been stated in this book or in [1]. After defining ψ_s, first C_d or depletion capacitance is calculated from the expression

$$C_D = \frac{dQ_s}{d\psi_s},$$
(5.1)

where Q_s is defined in Equation (4.3). Finally, n as a function of temperature at a gate voltage sufficiently enabling the subthreshold operation of n-MOSFET [5]:

$$n(T) = 1 + \frac{C_D(T)}{C_{ox}}$$
(5.2)

$$n(T) = 1 + \frac{\sqrt{\dfrac{\varepsilon_s \varepsilon_o q p(T)}{2\psi_s(T)}}}{C_{ox}} \left[1 + \frac{n_i^2}{p(T)^2} e^{\frac{q\psi_s}{kT}} - \frac{n_i^2}{2p(T)^2} \frac{kT}{q} \frac{e^{\frac{q\psi_s}{kT}}}{\psi_s(T)} \right],$$
(5.3)

where every term of Equations (5.2) and (5.3) have been defined earlier in relation to this research discussion. The simulation results of the surface band bending potential ψ_s as a function of gate to source voltage in the vicinity of subthreshold region operation of the 1 μm n-channel MOSFET with its specified device parameters are plotted below with substrate temperature variation from 100–500 K. As can been seen from Figure 5.1, the 100 K operated n-channel MOSFET exhibits a considerably higher subthreshold surface potential band bending value from a deep subthreshold state to almost on-state or close to threshold voltage state and the nonlinear increase at 100 K operated device characteristic is also sharper lending the observation that depletion width will be much wider first at deep subthreshold region and extend the widening state of the width even close to turn-on giving rise to very low subthreshold slope factor or the value itself that is a boon for the device professionals engaged with constant design and implementation of device features in connection to device architectural evolution to meet this bench mark criterion of reducing the subthreshold slope below 60 mV/decade. Compared to 100 K operation, operation of 200–500 K reveals surface band bending characteristics that are close in value and exhibit a less nonlinear increase from an extreme subthreshold to near threshold region operation of n-MOSFET. Therefore, based on the design needs, a substrate temperature engineering from 100–300 K should be sufficient to tune the n or subthreshold slope S to eventually control the switching transition or speed.

In conjunction to the above plot, the more relevant subthreshold slope factor reveals the following simulation outcome as a function of gate voltage near the vicinity of subthreshold region of operation at different substrate temperatures of operation.

As we can see from Figure 5.2, the subthreshold slope factor at 100 K operation exhibits a highly proportionately lower value from a deep subthreshold to near threshold compared to other temperatures operation 200–500 K. Since the surface potential at 100 K substrate temperature operation is fairly large, smooth progression from a typically higher value at deep subthreshold to lower value at near threshold is obtained. But for temperatures such as 200–500 K operation,

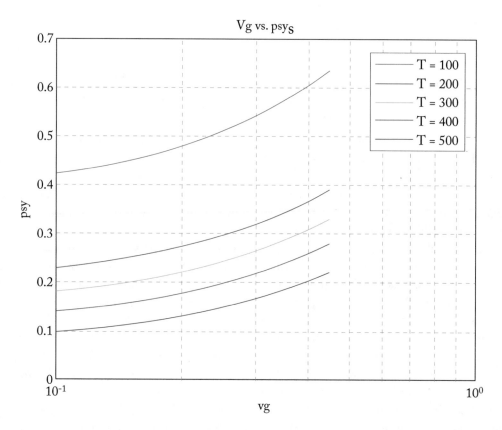

Figure 5.1: Subthreshold region surface band bending potential as a function of operated gate voltage for an n-channel 1 μm MOSFET at different substrate temperatures from 100–500 K.

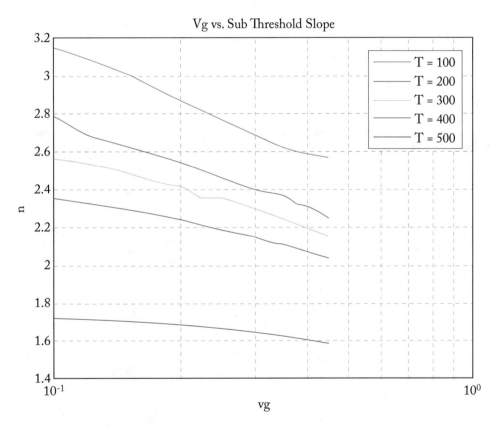

Figure 5.2: Subthreshold slope factor *n* as a function of gate voltage for a deep subthreshold to near threshold operation at different substrate temperature operations of *n*-MOSFET.

the surface potential decrease is much steeper and has a fluctuation characteristic in its value for gate voltages values in between extreme subthreshold to near threshold. This feature is a direct outcome of the more accurate modeling equation representation of n, as per Equation (5.3) where the added term in the third bracketed expression unveils this fluctuation nature of n due to surface potential band bending variation at these gate values. The depletion width or capacitance as a result undergoes variation in its values that are directly impacted by lower gate voltage defined fluctuations in surface band bending potential. Finally, Table 5.1 summarizes the extraction of subthreshold slope values at various substrate temperatures and justifies the selection of lower operating temperature to reap the benefit in the targeted values of subthreshold slope or subthreshold slope factor.

Table 5.1: Summary of the extraction of subthreshold slope factor values at various substrate temperatures

V_{gs} Value	n (T = 100 K)	n (T = 200 K)	n (T = 300 K)	n (T = 400 K)	n (T = 500 K)
0.1	1.7200	2.354	2.5661	2.782	3.15
0.125	1.707	2.3234	2.526	2.727	3.0817
0.150	1.6956	2.294	2.4877	2.6759	3.0089
O.175	1.6850	2.266	2.441	2.6322	2.9367
0.2	1.6743	2.2408	2.418	2.5855	2.87
0.225	1.6635	2.2166	2.358	2.542	2.822
0.25	1.654	2.1937	2.355	2.502	2.772
0.275	1.6447	2.1701	2.326	2.4662	2.727
0.3	1.635	2.1479	2.296	2.4304	2.6903
0.325	1.626	2.1286	2.271	2.4002	2.6540
0.35	1.618	2.089	2.247	2.3675	2.6270
0.375	1.6113	2.071	2.223	2.33	2.6034
0.4	1.6016	2.054	2.2001	2.311	2.5883
0.425	1.594	2.038	2.179	2.28	2.5788
0.45	1.5867	2.008	2.154	2.25	2.5780

CHAPTER 6

Review of Scaled Device Architectures for Their Feasibility To Low-Temperature Operation Simulation Perspectives of the Author's Current Research

To put the simulated outcomes of the previous chapters as narrated by the author into perspective where the readers and device engineering professionals can be knowledgeable of the efficacy of these simulated plots in terms of detailed understanding of today's scaled device architectures as batch fabricated in industry, the author draws the reader's attention to some well-cited and highly acclaimed published journal articles from literature survey. In this context, the author will first review the article by Ionescu and Riel [114] on Tunnel FET design, optimization, and performance analysis and during enumeration of the review, will also apply the concepts detailed in these articles and from the author's own device physics probed understanding to highlight the importance of the simulated outcomes provided in this book from the context of low-substrate temperature operational performance features of these industry-standard Tunnel FET architectures. The first review article on Tunnel FET as composed by Ionescu and Riel titled "Tunnel field-effect transistors as energy-efficient electronic switches" [114]. In this article, the author first discusses the importance of sub-60 mV/decade subthreshold slope attainability at room temperature (300 K) for an energy-efficient switch where carrier recovery for efficient switch turn-off can be triggered by band-to-band or quantum mechanical tunneling taking place of thermionic emission-based carrier injection and recovery. In modeling Equation (3) illustrated in [114], Ionescu and Riel draw the important analogy that total switching energy and hence power consumption is squarely dependent on supply voltage V_{DD} and linearly dependent on I_{off} (off-state leakage current) to I_{on} (ON-state drive current where the gate voltage overdrive is only several fractions up to 20% of 1 V above than TFET threshold voltage) ratio. Since Tunnel FET device structures are intrinsically scaled to operate at a lower supply voltage, hence very low V_{DD} and at the same time enables an I_{on}/I_{off} ratio close to 10^6, these devices do hold the immense potential for being ideal candidates for near-threshold logic switches benchmarked for extremely low-power dissipative switches. In the physics section of TFET in [114], the author first distinguishes the important carrier injection mechanism in TFET being inter-band tunneling whereby charge carriers transport in the case of n-TFET from one aligned energy band in the valence band into another aligned energy band in the conduction band of a degenerately doped p^+-n^+ junction at the source. The abruptness of turn-on or steepness of the tunnel current is therefore intimately governed by this interband tunneling mechanism that can be switched on and off abruptly by controlling the band bending; hence, band alignment for tunneling probability in the channel region by means of the gate bias. TFETs by virtue are ambipolar devices but by designing an asymmetry in the doping level or profile close to the surface of the p^+-n^+ source junction or by restricting the movement of the type of carrier that is not dominant type by using heterostructures, a tunneling barrier at the drain can be widened to suppress the ambipolarity. For a small range of supply gate voltage $V_G > 0$ in the case of n-TFET structure, the tunneling width or distance can be effectively reduced by the gate voltage by bringing the valence band of p^+ in close proximity of conduction band in n^+. In contrast to conventional n-MOSFET where the simulated outcomes presented by the current author revealed that subthreshold slope S declines with gate voltage in subthreshold to close to threshold range, [114] reveals that for TFET beyond a particular narrow range of V_G where reduction of S below 60 mV/decade is attainable, the S rather increases sharply with increasing gate voltage or for a substantial proportion of gate overdrive. This happens due to a decrease of inter-band tunneling with increasing gate voltage which puts the valence band and neighboring conduction band into misalignment for tunneling purposes. Ionescu and Riel [114] further make important observations that in order to boost the drive current of Tunnel FET, the transmission probability T_{WKB} resulting from inter-band tunneling has to be superiorly maximized. According to Equation (6) of [114], T_{WKB} is defined by an exponential function with a negative exponent containing product of m^* (tunneling mass), E_G (band gap), and λ the screening tunneling length. Of these three parameters, selecting materials with low enough band gap material at the source p^+-n^+ tunneling junction will reduce E_G and the $\Delta\phi$ (the energetic difference between valence band at source and the conduction band at the channel constituting the tunneling volume) and will strongly impact T_{WKB}. By contrast, λ and m^* cannot be modulated over a wide range to impact T_{WKB}. Therefore for today's device design technology for TFET structure after suitably defining the heterojunction at the source to tune E_G and $\Delta\phi$, engineering technique is applied to reduce λ for a small gate voltage range and utilize transverse plane of transport in addition to strain engineering to effectively reduce m^*. The value or significance of this Equation (6) for T_{WKB} can further be extended to the low-substrate temperature operation of the TFET from the current author's perspective. At a low substrate temperature for group IV materials like silicon and germanium, both E_G, $\Delta\phi$, and m^* are significantly enhanced at sufficiently reduced substrate temperatures of operation leading to attenuation of T_{WKB}. Hence, additional device engineering techniques must need to be applied at the p^+-n^+ tunneling junction to reduce λ and control the extension of E_G, $\Delta\phi$, and m^* so that these values are within a tolerable margin of the values reported for room temperature operation. Although the overall tunneling probability of carrier injection at the source junction is reduced at lower operating temperature, the drift or transit time through the channel for the carriers is reduced significantly due to much enhanced carrier mobility at the operated reduced crystal temperature and when these carriers are collected from further reverse biased drain contact, the higher drift velocity of carriers partially nullifies the reduced tunneling probability at the source and as will be further reported in the review, the overall on-state drain current for a low enough gate voltage at the substrate temperature range 100–200 K is nearly in-line with the value reported at room temperature or 300 K [76]. In this reviewed article the author further comments that in order to boost T_{WKB} the screening tunneling length λ should be effectively minimized by ingenious design of device geometry at the source including dual or triple work function-tuned metal gates, doping profile optimization close to junction surface and gate capacitance increase by selection of proper high-k material, reduced dielectric thickness, and gate-sidewall spacer control by overlapped gate to p^+-n^+ source junction. In addition, body thickness t_{si} of the channel must be minimized for increased gate to body coupling. Embedding a highly doped n^+ pocket layer close to p^+ abrupt junction will further make the band steeper with band gap narrowing effect essentially resulting in decreased tunneling distance λ. The on-state current can be further enhanced by choosing heterojunction at the source with a smaller E_G for the source emitter diode and higher E_G for the drain junction can be adopted to minimize ambipolar transport impacting I_{off}. The author further elaborates that a combination of steeper subthreshold slope S and higher I_{on} can be achieved with moderate doping of n^+ layer of the source junction and staggered or broken gap line up. By judicious control of overlapped gate with source tunneling junction or designing a source region covered with epitaxial intrinsic channel layer under the top gate extension, I_{on} can be improved by more than a factor of 10 and a low S_{avg} (average subthreshold slope) can be obtained. Since dopant activation and incomplete ionization are impeding artifacts at lower substrate temperature operation of TFET, highly degenerate doping is required for the above-discussed doping profile engineering techniques. The other potent ways to optimize TFET device performance when lower substrate temperature is chosen are device geometry adaptation of the source, i.e., choosing multiple gate layer with gate work function optimization which will not infringe with device temperature, dielectric thickness and permittivity not affected by device temperature and gate overlap extension to source hardly affected by device temperature. The article claimed that TFET structures fabricated on $Si_{1-x}Ge_xOI$ substrate ($x = 0$, 15%, 30%) has a measurement I_{on} 335 fold increase for n-TFET compared to pure SOI ($x = 0$) substrate and 2700 fold increase of I_{on} for p-TFET compared to pure SOI substrate. By replacing group IV materials like Si and Si:Ge source with group III-V materials like InAs and InSb, I_{on} increases by several orders of magnitude and can be reached at lower reverse bias field of the source junction and this has a positive impact for low-temperature operation of TFET as apart from controlling band gap engineering to offset the band gap increase associated with lower crystal temperature operation, low reverse fields exist at the tunneling junction for these substrate temperatures due to dopant activation retardation taking place for both p^+ and n^+ junction at the source. A rather powerful implication of using TFET at lower substrate temperature operation comes out from the proven fact that with high-k gate dielectric, gate coupling to source tunneling junction is strongly enhanced and since T_{WKB} is exponentially impacted by dielectric permittivity, the end result is that I_{on} is significantly increased and additional slight increase of I_{off} will not make any influence on already low I_{off} for all TFET structures [76]. Another highly important feature that comes with operating TFET devices at lower substrate temperature is that I_{GIDL} (gate-induced drain leakage current) dominated by band-to-band tunneling for small negative gate voltage V_G is reduced substantially at these lower substrate operating temperatures as the BTBT field is less steep and abrupt due to lower p^+-n^+ source junction dopant activation in addition to increased band gap of the junction reducing BTBT tunneling effect. I_{GIDL} is an important component of off-state leakage current for digital circuits such as inverters and memory elements and the minimization of I_{GIDL} is extremely important for TFET structures as these devices tend to suffer from increased level of I_{GIDL} for low negative gate voltage values due to BTBT induced reverse tunneling effect at the reverse biased drain (n^+) to channel junction. There are other off-state leakage current components in n-TFET devices generated by trap assisted tunneling (TAT) mechanism and SRH (Shockley–Read–Hall) generation-recombination process that exists in a reverse biased highly doped p-n junction and both these components provide more variation to I_{GIDL} at elevated substrate temperatures higher than 300 K compared to lower substrate temperature operation.

6.1 SILICON NANOWIRE TRANSISTOR PERFORMANCE ANALYSIS WITH CONSIDERATION OF LOW-TEMPERATURE OPERATION

After analyzing the Tunnel-FET performance features from the perspective of their beneficial suitability to lower temperature operation a theme for this book, this sub-section draws the reader's attention to the review of the classic paper written by Lu et al. [98]. From the excerpts of the signature abstract of the paper it is evident that the drive toward scaling sustained nanowire architecture in the 10 nm node should focus on nanowire heterostructure integration or fabrication and thereby giving way to performance attainment near ballistic limit and exceeding state-of-the-art planar devices. In the introduction section of [98], while solving the surface potential by one-dimensional Poisson's equation, the author thus reveals the characteristic length λ for the nanowire material which is the determinant metric about how efficiently the gate controls the body potential by the term "gate integrity" essential for gate capacitance enhancement, superior channel inversion, and superior off-state leakage current as the scaling continues progressively lowering the maximum gate and drain voltage that can be applied. λ for a one-dimensional structure has been known to be $\sqrt{\frac{\varepsilon_{si}}{\varepsilon_{ox}} t_{si} t_{ox}}$ and therefore for better gate integrity where the channel surface potential is more controlled by gate rather than drain field penetration, λ needs to be reduced enabled by downscaling of silicon body thickness t_{si} and oxide thickness t_{ox} simultaneously while upscaling the term ε_{ox} by choosing high-k gate dielectrics. Furthermore, instead of two-dimensional planar nanowire, a surrounding gate or wrap-arround gate nanowire (NW) structure in cylindrical assembly reduces λ more that is not intrinsically possible by the term parameters mentioned above. The author further comments on the synthesis and fabrication of these NW devices that the nanocluster-catalyzed vapor-liquid-solid (VLS) growth process, particularly for CVD (chemical vapor deposition) or CBE (chemical beam epitaxy) techniques, offers the ability to fine tune the diameter (t_{si}), morphology, and electrical properties of NWs in a flexible and controllable fashion. The author comments from the perspective of ULSI manufacturability that the feasibility to control nanowire growth down to the atomic level is one of the main factors leading to the great success that nanowire research enjoys today. During elucidation of diverse nanowire growth techniques employed for NW device characterization and electrical parametric measurements, the authors of [98] reveal the universal bottleneck associated with atomically sized nanowire pores being source and drain contact resistances. All suitably chosen metal S/D contacts regardless of metal sources on degenerately doped source and drain junctions end at providing Schottky barrier contacts rather than intended ohmic contacts. The authors confer that positive Schottky barriers are observed at the metal/S-D junction interface due to combined effect of metal work function and Fermi level pinning by surface states. Cui et al. [115] researched source-drain contact thermal annealing and surface passivation on p-type silicon nanowire and reported an average transconductance increase from 45 nS to 800 nS on passivated NW devices and average mobility upsurge from 30–560 cm^2/V-s on passivated devices. The annealing process for these devices is worth mentioning and the annealing treatment was carried out at 300–600°C for 3 min in the forming gas (10% H$_2$ in He) to improve the contact ohmic nature and passivate Si-SiO$_x$ interface traps. Cui et al. [115] reviewed this connection from the context of its impact of annealing on S/D contact resistance the author used Ti as the source/drain contact metal giving rise to TiSi$_2$ after saliciding process and sintering with a low effective Schottky barrier height on p-type silicon. Other transition and rare earth metal like La and Mo metal complex and La and Mo silicides usually modify and lower the Schottky barrier upon contact by imparting additional band tail states in the gap formed by the Schottky barrier and hence reduces contact resistance. The Schottky barriers formed on the naturally grown contacts on S/D junctions may function as tunnel barriers when both the metal Fermi level and S/D material Fermi levels are based on degenerate doping and have a pejorative consequence to the operation of these devices at lower substrate temperatures than 300 K. This feature is further discussed by the authors of [98] who discussed that at lower temperature additional transport features namely Coulomb blockade (CB) and coherent charge transport that are absent at room temperature are conspicuous at these lowered temperatures of operation down to 5 K with pore size between 3–6 nm. These devices exhibit periodic $I_{ds} - V_{gs}$ curve oscillations at these lowered substrate temperatures of operation where quantum dot is formed by the entire length of nanowire between S/D contacts. The author later discusses that as an alternative to silicon NW, germanium NW FET has been synthesized using the nanocluster-catalyzed VLS approach owing to germanium's higher electron and hole mobility. In addition, Ge has a reduced band gap than Si which results in lower Schottky barrier height when metal contacts are made on Ge source/drain. Despite presence of additional surface defect states in Ge compared to silicon that contribute to Fermi level pinning in Ge, its impact on much reduced E_G of Ge and thus φ_B of Schottky barrier is much less compared to metal-SiNW contacts. As another alternative to SiNW for today's scaled devices, the author further discusses InAs nanowire material prized with its small effective electron mass ($0.023m_o$) resulting in higher electron mobility in InAs bulk smaple. In addition, an electron gas layer is known to form at the surface of planar InAs due to Fermi level pinning in the conduction band at the surface. Furthermore, the formation of an electron gas combined with InAs's small band gap (0.35 eV) should relatively yield transparent contacts to InAs nanowire substrates. The authors systematically usher in the need of the heterostructured NW devices since transparent ohmic nature contacts which remain challenging from fabrication standpoint for degenerately doped S/D with rather intrinsic or moderately doped channel substrate, can be suitably achieved in heterostructured NW devices by modulating the band structure at the conduction or transport surface particularly at the gate overlapped source-channel junction. The authors of this paper [98] put emphasis on radial and axial nanostructure with Ge/Si core/shell system radial NW in which a 3–5 nm Si shell is epitaxially grown on top of a 10 nm diameter Ge nanowire core. Due to the large (0.5 eV) valence band offset between Ge and Si, the Fermi level pinned inside the Si band gap is below the Ge valence band giving a negative Schottky barrier and acting as a reservoir for free carriers sink and source. Also, metal-induced gap states or surface states that act as scattering centers for hole injection from this type of device are absent in this structure giving rise to higher contact induced source injection velocity of holes in this structural assembly. Compared to homogeneous SiNW which cause oscillations in drain current at lower temperature, these heterostructured Ge/Si core/shell p-NW devices show contact characteristics that remain transparent even at low temperatures when the device is fully depleted (thin t_{si}). An n-channel radial heterostructured NW device that can be conveniently achieved in an assembly of undoped GaN/AlN/AlGaN show the formation of electron gas inside GaN core. Precise control of both the shell thickness and composition during MOCVD nanowire growth process allow fine tuning of the band structure and contribute to the atomically sharp interface between the core and shell layers. A 2 nm AlN dielectric layer sandwitched between GaN/AlGaN imparts larger conduction band discontinuity for better confinement of electrons and to reduce alloy scattering from the AlGaN outershell subjected to the fact that at lower operating temperatures, modulation-doped heterostructures of all types of NW devices suffer from alloy scattering that reduce peak carrier mobility and impact device transit time of carries or intrinsic speed. The authors in this paper further state that transport measurements carried out on GaN/AlN/AlGaN heterostructured nanowires exhibit intrinsic electron mobilities of 3100 cm²/V-s at 300 K and 21,000 cm²/V-s at 5 K. A near ideal subthreshold slope of 68 mV/decade at $T = 300$ K and $I_{on}/I_{off} = 10^7$ were obtained in a measurement setup as reported in this article. In a paper of Rastagi et al. [77], the authors made a fundamental analysis of nanowire drain current Vs. gate voltage transfer curves at reduced substrate temperature operation. Through experimentally fabricated NW device characterization, the authors of this paper prove by experimental demonstration that contrary to the standard observation of the increase of drive current of conventional n-MOSFETs at reduced substrate temperature at low drain voltage V_{ds}, the fabricated NW devices had achieved lower drive current at the reduced value of V_{ds} for reduced temperatures up to 200 K. A key fact of low-temperature transport in these NW devices is the occurrence of rapid and intense inter sub-band scattering masking the usual reduced phonon scattering factor on account of developed enhanced sub-band splitting as a result of increased quantum confinement effects at these temperatures. Since the separation of split-subbands comparatively widen at lower T, occupancy of these bands by carriers at lower V_{ds} is less probable and hence even at inversion, the current magnitude declines. As V_{ds} is increased and operating temperatures are 200 K or above, I_{ds} increases with temperature due to carriers populating the intersubband valleys with their separation decreasing with increasing V_{ds}. This observed fact showers critical implications for NW devices operated at low temperature particularly for low drain and gate voltages required from the context of scaling. Since the enhancement of drive current in NW devices for reduced T is only feasible with larger V_{ds} value, optimization of nanowire pore radius (t_{si}) by relaxing it will subsequently lessen quantum confinement effect at lower operating temperatures and help to reduce V_{ds} value. The other way is to control sub-band splitting by reducing the spatial separation of intersubband valleys subject to lower temperature operation by ingenious band engineered heterostructured radial and axial NW devices which have been discussed in the classic paper reviewed in this sub-section. Finally, with regard to GIDL current in the off-state of nanowire transistors, the author would like to draw the attention of the readers to the reference [116] where a comprehensive physical insights have been put to analysis by the authors in properly profiling GIDL current in nanowire FETs for V_{gs} value up to -1.5 V and wire diameter 5 nm. Added triggering mechanism results from an intrinsic bipolar transistor-(BJT) based positive regenerative feedback where source n^+ (emitter) and drain n^+ collector get positive feedback from the channel region (p-type base) which gets accumulated by holes for the off-state gate voltage bias $V_{gs} < 0$. The GIDL current in silicon NWFETs as per the author's observation in this reference article, can be minimized by lightly doped source and drain extension regions so as to modify the tunneling distance or width over which I_{GIDL} depends and is significantly controlled. For on-state current I_{on}, the lightly doped source and drain extensions do contribute to source and drain series or access-resistance enhancements lowering the maximum I_{on}, enhancing the potential at the intrinsic node of the source extension edge and therefore for the same gate voltage, reducing V_{gs} resulting in lower source to channel injection velocity. As a result, I_{on} will be reduced mostly by additional series resistances of source and drain extension regions. Therefore careful control of the extent of the source and drain lightly doped extension regions to effectively set the source and drain extension region series resistances is reported in this article as a state-of-the art modeling parameter to impact I_{GIDL}, I_{on}, and I_{on}/I_{off} ratio. From the context of lower substrate temperature operation of this nanowire FETs with lightly doped source and drain extension regions, since both n^+ source and n^- source extensions will be affected by incomplete ionization and band gap enhancement and this properties will also be visible for n^+ drain and n^- drain extension regions. As a result, I_{GIDL} will decrease at lower substrate temperature as the tunneling width will be larger at both source and drain extension regions aided by less steep energy bands. Although I_{GIDL} decreases, simultaneously I_{on} will decrease at on-state due to incomplete ionization of a source region extension lightly doped junction, the series resistance will be larger than room temperature case, a result of which V_{gs} for a given gate voltage will be much smaller compared to a room temperature case. For reduced V_{gs}, the carrier injection velocity at the source extension edge will be subsided and the on-current thus will be attenuated. This observation again enforces the point that both lightly doped source and drain extensions length and the actual lightly doped doping profiles and values are pivotal modeling parameters.

6.2   NEGATIVE CAPACITANCE FERROELECTRIC FET (NCFET) PERFORMANCE ANALYSIS WITH CONSIDERATION OF LOW-TEMPERATURE OPERATION

The third exploratory device architecture apart from Tunnel-FET and nanowire-FET which has received sizable attention from industry and academia is the negative capacitance ferroelectric FET (NCFET). Exploiting the characteristics of the Curie temperature (T_C) and the polarization loop of a ferroelectric dielectric sandwitched between top gate electrode, internal metal electrode, and underlying conventional SiO_2, the overall gate capacitance can be tuned to be less than zero (< 0) for certain gate voltage range and hence provides the capability of bringing down the subthreshold slope less than 60 mV/decade—a thermal transport based limit at room temperature. In addition while the dielectric permittivity being impacted by the polarization loop is modulated such that total gate capacitance turns out negative boosting the surface potential during inversion condition for a range of gate voltage providing internal voltage amplification desirable from inducing higher inversion charge density at the channel surface and contributing to on-current enhancement. In this sub-section, owing to the growing importance of study of NCFET from the scaled device performance improvement conforming to ITRS projections, the author puts forth the review and critical analysis of several highly cited classic papers focusing on ferroelectric FET modeling and performance assessment from the perspective of Moore's Law based scaling. The first paper reviewed in this context is Lee et al. [117]. The transistor model used for simulation in this paper is the gate-electrode/ferroelectric layer ($BaTiO_3$)/internal electrode/SiO_2/p-type Si substrate. The SiO_2 thickness was chosen to be 1 nm for the simulation purpose to highlight the need of aggressive scaling node of 10 nm and below. From the simulation outcomes performed by the authors in Figure 3(a), it became obvious that the slope $d\psi_s/dV_G$, a determinant for steeper subthreshold slope, becomes steeper as the ferroelectric thickness T_{FE} is increased. Their simulation results also confirmed that while T_{FE} of 450 nm has a huge internal voltage gain resulting in super steep switching feature of NCFET, a wide hysteresis window of 130 mV exists as a trade-off. From the Landau–Khalatnikov (L-K) equation used to determine the ferroelectric gate voltage V_{FE} as a function of dielectric polarization P, the nonlinear term that contributes most to the hysteresis window widening is $\beta(-1 \times 10^9$ m^5/F/coul^2 for BaTiO$_3$). A larger β will enhance the ferroelectric hysteresis window and induce a time lag between charge injection (forward sweep) and charge decay (reverse sweep) occurring at the source-channel junction during on-set of subthreshold region conduction. Therefore, to implement super-steep switching NCFET device at room temperature with sufficient polarization, the ferroelectric thickness layer T_{FE} must be optimized to control hysteresis window expansion. In Figure 3(b) of [117], the authors reveal another illuminating simulation outcome that with increase of body or channel doping N_A, the relative slope $d\psi_s/dV_G$ enhances when larger channel doping is selected subsequently enabling steeper subthreshold slope. This is contradictory to the fact that conventional MOSFET delivers steeper subthreshold slope at 300 K when the channel doping is low or near intrinsic value. Due to the larger polarization enhanced non-linearity in ψ_s and internal ferroelectric potential V_{FE}, the polarization or hysteresis window is much wider when the channel doping is higher. Their simulations revealed that for N_A doping of 10^{16}/cm^3, 10^{17}/cm^3, and 10^{18}/cm^3, the hysteresis window width extracted was 20 mV, 40 mV, and 70 mV. From the low-temperature operation perspective and even at room temperature operation of scaled MOSFETs, the scaled performance of device architectures improves when the channel doping is maintained near intrinsic level. Therefore, according to the results of Figure 3(b), if the channel body is optimized to reflect the intrinsic level value, the device will exhibit slightly reduced steepness of subthreshold slope but at the same time polarization window will be narrower speeding up both charge injection process and the removal process at the source-channel junction during subthreshold conduction state. In Figure 4 of [117], the authors clearly demonstrate through TCAD simulation that the ferroelectric MOSFET provides superior on-current compared to conventional MOSFET and also the steeper value of current-voltage slope for ferroelectric FET is readily visible from Figure 4. Finally, at the conclusion of this article write-up, the authors comment that in order to engineer both steeper subthreshold slope and narrower hysteresis window, the best optimization will be choosing thicker ferroelectric layer (T_{FE} higher, higher subthreshold slope) and compensating the attendant increase of polarization window with lower channel doping N_A.

In addition, the industry choice of ferroelectric material compatible for ULSI fabrication such as widely researched PbZrTiO$_3$ (PZT) and BaTiO$_3$ and proper modulation of the ferroelectric material parameters (α, β, γ, and ρ) in terms of definable values of remnant polarization P_r and coercive voltage V_c arising out of experimentally calibrated polarization-field loop near the Curie temperature will also play a role to the extent $d\psi_s/dV_G$ can be controlled by other circuit and device parameters in conjunction with these material selective parameters. A paper by Salahuddin et al. [118] studies the impact of parasitic source and drain capacitances, the value of remnant polarization P_r, and coercive field E_c on NCFET internal amplification. Their findings corroborate the fact that with increase of E_c (wider polarization window) DC gain is enhanced for a small gate voltage range. Also with the decrease of P_r (less steep polarization window) DC gain is enhanced for a small gate voltage range. In another paper by Shin et al. [119] the authors propose another way of controlling the hysteresis window of ferroelectric gate dielectric by utilizing the extension length L_{ext} controlled source and drain doping extension regions encroaching into the channel and subsequently decreasing the spacer controlled C_{gs} and C_{gd} (gate-to-source capacitance and gate-to-drain capacitance) delimiting the value of ψ_s and as a result of which V_{FE} decreases and hysteresis window narrows. Importantly the authors in this paper used PZT ferroelectric material which is more widely researched as the staple ferroelectric dielectric material by both academicians and industry professionals. From the experiment the authors of the article confirm that the hysteresis window rapidly decreased from 1.02 V to 0.48 V as L_{ext} increased from 60 nm to 150 nm. The only drawback for this design adaptation is the associated increase of source and drain series resistances resulting in a decrease of maximum achievable on-current and slightly enhanced lay-out area which can be problem from lay-out constraints defined by ITRS projection for scaled nodes. The augmentation of L_{ext} for NCFET according to this paper was from 60–150 nm, although decreased hysteresis window as per the above quoted values increased the average subthreshold slope S_{avg} from 10 mV/decade to 20 mV/decade for reverse sweep and 20 mV/decade to 80 mV/decade for forward sweep. Therefore, when the devices switch from relatively off-state to on-state which is the case for forward sweep, the subthreshold slope is more critically worsened by the incorporation of source and drain extension regions. In Khan et al. [102], the temperature impact on ferroelectric FET is studied for PZT-STO (SrTiO$_3$) bilayer. The ferroelectric capacitance C_{FE}, dielectric capacitance C_{DE} and total gate capacitance $C_{FE} + C_{DE}$ are all enhanced at temperatures up to 200°C close to Curie temperature. Hence, in the intended low temperature operation of the scaled device architectures, when the temperatures are in the range of 100–200 K, for the currently selected ferroelectric materials such as PZT, BaTiO$_3$, and STO, sufficient polarization will not take place to properly boost the C_{FE} and $C_{FE} + C_{DE}$ capacitances and hence impart strong negative capacitance (NC) effect to device internal voltage gain. But low temperature operation would still suffice for these NCFET devices if the temperature is not too low beyond room temperature but kept and managed between 250–280 K. Operating the device slightly below room temperature will not appreciably alter the polarization process defined P_r and E_c values obtained at room temperature of operation. The authors of this paper [103] also noted that higher temperature operation than room temperature of the NCFET essentially makes the ferroelectric loop more steeper and width more narrower. As a result of operation of these devices higher than room temperature, the coercive field E_c decreases and the remnant polarization P_r increases and in the analysis of the here-to-forth review it was clarified that DC internal amplification results for higher E_C and reduced P_r values of the polarization loop. All these temperature related simulation outcomes enforce the observation that with the tuning of various ferroelectric material related parameters including its thickness, the device performance can be conveniently obtained to enable ITRS projection at room temperature or 300 K operation with room of improvement resulting from slightly below 300 K operation, i.e., 250–280 K where the device speed at the on-region can be further increased by boosting predominantly phonon-limited and Coulomb scattering-related channel carrier mobility even in the midst of polar and non-polar optical phonon scattering and piezoelectric scattering induced transport characteristics due to the presence of ferroelectric dielectric material at low drift field and high drift-field. Additionally, the subthreshold slope factor will proportionately reduce with substrate temperature (slightly reduced than room temperature) aided by the multiplicative negative capacitance-induced reduction factor.

CHAPTER 7

Summary of Research Results and Conclusions

The previous book composed by the present author ushered in a novel concept as a panacea to device scaling bottlenecks: the current device architectures of n-MOSFETs are constantly encountering ITRS projection-based device performance benchmarks and introduced the potent concept that operating the scaled device architectures including the currently fabricated nanowire devices under lattice or substrate temperatures of 100–300 K, almost all benchmark figures as set by ITRS can be conveniently adhered to. The previous book in this regard provided simulation results of threshold voltage variation for a 1 μm channel length n-MOSFET device as a function of substrate temperature and further projected the outlook of threshold voltage variation window from the scaled supply voltage reduction perspective of each scaling node. The book also showed a method of on-die temperature control based on device partitioning scheme of different logic based systems and regulating die temperatures of these logic blocks by microcontroller controlled solid state miniature refrigerators as per IC-based TCAD solutions of threshold voltage control requirements from the device speed perspective, required on-current perspective , required quiescent power dissipation perspective and total off-state leakage power dissipation perspective as analyzed for different ASIC based digital logic blocks fabricated on a device-partitioning-based scheme on the die. The current book has augmented the applications of lower substrate temperature operations of today's scaled node configuring the Moore's Law-based evolutional era by simulation works on on-state drain current, subthreshold drain current, channel mobility, and subthreshold slope such that each of these parameters is superiorly improved at lower operating temperature operation of n-MOSFETs and hence the revelation through device analytical simulation results of occurrences of these much needed performance benefits serve as a denouement of the intricate infeasibility posed by attendant non-scalability of today's 10 nm MOSFET architectural derivatives to aid in improvement of these parameters by sole operation at room temperature or 300 K. Different device architectures manufactured today preserving the scaled evolution of nodes fail to achieve either the targeted maximum on current or the lowest value of off-state leakage current or bolstering the channel mobility what stress, strain, wafer orientation technology, and metal gate with high-k dielectric architecture cannot achieve or provide a sufficiently low subthreshold swing in the vicinity of 30 mV/decade or lower. Except Tunnel-FET and Ferroelectric FET architectures, most device architectures at the current technology node of Moore's Law-based scaling are centered on the concept of improving gate integrity control of the channel with channel doping close to intrinsic level. However, these features maintain the required on-state drain current and off-state drain current as per ITRS roadmap, they are inadequate to provide the required on-state channel mobility and steep subthreshold slope or lower subthreshold swing, probably one of the very important benchmark figure-of-merit from energy switching based power reduction for subthreshold logic devices where the required power supply of operation is in the vicinity of 0.1–0.2 V. Tunnel-FET has received much attention these days to welcoming device engineers for its ability to provide subthreshold swing less than 60 mV/decade as set by room temperature thermal energy limit. But since the transport of electrons in these devices is principally governed by source-to-intrinsic body tunneling, this results in lower on-state current drivability and hence will not be useful for intense switching and computation based logic architectures like ring oscillators, arithmetic logic units (ALU) and other elements of microprocessors or microcontrollers and their applications thus will have to be in subthreshold logic based devices and extremely low quiescent current analog devices such as current sources and DC-DC voltage regulators on the die in mV range voltage boosting. Because of moderate flow of electrons from the source junction of Tunnel-FET due to band-to-band tunneling, drift mobility of intrinsic layer is not significant to trigger a boost in the on-state drive current at room temperature. The only other benchmark figure-of-merit where Tunnel-FETs contribute is the off-state leakage current I_{off} which is the lowest of all the device architectures currently fabricated including gate integrity-based device architectures. From the perspective of feasibility of Tunnel-FET devices at lower substrate temperature operation, tunneling will be reduced at the source junction due to band gap enhancement at the source due to reduced substrate temperature of the device but the drift speed through the intrinsic layer can be significantly enhanced by reduced operating temperature and eventually the on-state current will not diminish solely as a result of substrate temperature reduction rather will be neutralized to remain almost at par with room temperature Tunnel-FET drive current value. Ferroelectric FET is also an important candidate to pursue at the end of Moore's Law due to the reason that Ferroelectric FETs achieve less than 60 mV/decade subthreshold swing at room temperature by the ferroelectric effect of the ferroelectric dielectric of the gate resulting in a negative capacitance and reduction of subthreshold swing. The Ferroelectric FETs are also capable of internal amplification of gate field thus create better inversion carrier density underneath the gate resulting in higher on-current. But the off-state leakage current of ferroelectric FET will not be as low as enable by gate integrity controlling based other device structures and Tunnel-FET devices. Additionally, channel mobility is not enhanced to the level required from circuit speed for the scaled nodes when operational scenarios of these FET devices at room temperature is considered. Controlling Curie-temperature and Ferroelectric loops of these FET structures is also important for proper gate bias range and drain voltage bias range during on-state operation and also the fact that both of these voltages at the gate and drain reach mV range where sufficient magnetization or ferroelectricity of the gate dielectric might be a challenge to obtain as per scaling node device structure. For the operation of these devices at lower than room temperature, if the ferroelectric loop is still not degraded much, the subthreshold swing can be reduced from (i) negative gate capacitance effect and (ii) direct temperature-based scaled reduction effect. In addition, when the proper gate bias is chosen to provide internal gate-to-channel field amplification, the final on-current drive will be further enhanced as at reduced substrate temperature, inversion channel electrons move with a faster channel mobility. A review and critical analysis of current potential and trendsetting research articles on Tunnel-FET, gate-all-around nanowire FET, and ferroelectric negative capacitance FET elaborated for the readers' perusal in Chapter 6 confirm the fact that all these devices will be aided by lower off-current I_{off} at reduced substrate temperatures (<300 K) and the next positive impact will be on subthreshold slope reduction. Unfortunately, the device on-current I_{on} enhancement conforming to the operated scaled node cannot be guaranteed for Tunnel-FET, nanowire FET, and ferroelectric FET as the natural enhancement of channel mobility as is the case for conventional MOSFET is not extractable from reduced substrate temperature operation for the aforementioned device architectures. For tunnel-FET another potent I_{on} reduction factor at lowered substrate temperature is the widening of the bandgap width for tunneling at the gate-to-source junction. Notwithstanding the I_{on}/I_{off} ratio still boosted for all the aforementioned device architectures currently being researched and manufactured owing to the reason that for little change of I_{on} at reduced temperature operation, a benefit accrued from reduced I_{off} is substantial.

The author concludes this book with the observation that since almost all performance benchmark figure-of-merits of a scaled MOSFET device are improved by operating these devices at lower substrate temperature, the device engineering professionals and researchers can now devote their time and research for implementing the concepts and illustrations narrated in this book to demonstrate the fact that Moore's Law lives at least to 1 nm technology node if we are ready to make accessory devices to supplement the low-temperature operational needs of these devices co-fabricated on-chip and provide higher yield and reliability-based operational devices to the supply markets and end-users.

References

[1] Ashraf, N. S., Alam, S., and Alam, M., *New Prospects of Integrating Low Substrate Temperatures, with Scaling-sustained Device Architectural Innovation*, Synthesis Lectures on Emerging Engineering Technologies, Lecture #4, Morgan & Claypool Inc., pages 1–80, 2016. DOI: 10.2200/s00696ed1v01y201601eet004. vi, 1, 2, 7, 9, 10, 11, 16, 23, 36, 38, 39, 44

[2] Pierret, R. F., *Semiconductor Device Fundamentals*, pages 23–74 and 563–690, Addison-Wesley, Reading, MA, 1996. 19, 20, 21

[3] Pierret, R. F., *Advanced Semiconductor Fundamentals, Modular Series on Solid State Devices*, Pearson Education Inc., Upper Saddle River, NJ, pages 87–133 and 175–215, 2003. 19

[4] Streetman, B. G. and Banerjee, S. K., *Solid State Electronic Devices*, 6th ed., Pearson Prentice Hall, Upper Saddle River, NJ, 2006.

[5] Taur, Y. and Ning, Tak H., *Fundamentals of Modern VLSI Devices*, pages 112–221, Cambridge University Press, 1998. 26, 27, 35, 36, 37, 38, 45

[6] Sze, S. M., *VLSI Technology*, McGraw Hill, NY, 1983.

[7] Sze, S. M. and Ng, Kwok K., *Physics of Semiconductor Devices*, John Wiley & Sons Inc., NJ, 2007. DOI: 10.1002/0470068329.

[8] Sze, S. M., *Semiconductor Devices: Physics and Technology*, John Wiley & Sons Inc., NJ, 1985.

[9] Shur, M., *Physics of Semiconductor Devices*, 1st ed., Prentice and Hall, 1990. DOI: 10.1063/1.2810727.

[10] Shur, M., *Introduction to Electronic Devices*, John Wiley & Sons Inc., NJ, 1996.

[11] Neamen, D. A., *Semiconductor Physics and Devices: Basic Principles*, pages 106–191 and 371–490, 4th ed., McGraw Hill Inc., NY.

[12] Tsividis, Y. P., *Operation and Modeling of the MOS Transistor*, pages 35–216, McGraw Hill International Editions, 1988.

[13] E-book on *Principles of Semiconductor Devices*, Bart Zeghbroeck.

[14] Colinge, J. P. and Colinge, C. A., *Physics of Semiconductor Devices*, Springer Science and Business Media Inc., 2002. DOI: 10.1007/b117561.

[15] Schroder, D. K., *Advanced MOS Devices, Modular Series on Solid State Devices*, vol. VII, Addison-Wesley, Reading, MA, 1987.

[16] Barber, H. D., Effective mass and intrinsic concentration in silicon, *Solid State Electronics*, vol. 10, pages 1039–1051, Pergamon Press, 1967.

[17] Vankemmel, R., Scchoenmaker, W., and Meyer, K. De., A unified wide temperature range model for the energy gap, the effective carrier mass and intrinsic concentration in silicon, *Solid State Electronics*, vol. 36, no. 10, pages 1379–1384, 1993. DOI: 10.1016/0038-1101(93)90046-s.

[18] Varshni, Y. P., Temperature dependence of the energy gap in semiconductors, *Physica*, vol. 34, no. 1, pages 149–154, 1967. DOI: 10.1016/0031-8914(67)90062-6.

[19] O'Donnell, K. P. and Chen, X., Temperature dependence of semiconductor band gaps, *Applied Physics Letters*, vol. 58, no. 25, pages 2924–2926, 1991. DOI: 10.1063/1.104723. 9

[20] Taur, Y., Buchanan, D. A., Chen, W., Frank, D. J., Ismail, K. E., Lo, S.-H., Sai-Halasz, G. A., Viswanathan, R. G., Wann Hsing-Jen, C., Weind, S. J., and Wong, H.-S., CMOS scaling into the nanometer regime, *Proc. of the IEEE*, vol. 85, no. 4, pages 486–504, 1997. DOI: 10.1109/5.573737. 1, 2, 9

[21] Frank, D. J., Dennard, R. J., Nowak, E., Solomon, P. M, Taur, Y., and Wong, H-S., Device scaling limits of Si MOSFETs and their application dependencies, *Proc. of the IEEE*, vol. 89, no. 3, pages 259–288, 2003. DOI: 10.1109/5.915374.

[22] Frank, D. J. and Taur, Y., Design considerations for CMOS near the limits of scaling, *Solid State Electronics*, vol. 46, pages 315–320, 2002. DOI: 10.1016/s0038-1101(01)00102-2.

[23] Wong H.-S. P., Frank, D. J., Solomon, P. M., Wann, C. H. J., and Welser, J. J., Nanoscale CMOS, *Proc. of the IEEE*, vol. 87, no. 4, pages 537–570, 1999. DOI: 10.1109/5.752515.

[24] Wong Hon-Sum, P., Beyond the conventional transistor, *IBM Journal on Research and Development*, vol. 46, no. 2/3, pages 133–167, 2002. DOI: 10.1147/rd.462.0133.

[25] Taur, Y., CMOS design near the limit of scaling, *IBM Journal on Research and Development*, vol. 46, no. 2/3, pages 213–222, 2002. DOI: 10.1147/rd.462.0213.

[26] Dennard, R. H., Gaensslen, F. H., Yu, Hwa-Nien, Rideout, V. L., Bassous, E., and Leblanc, A. R., Design of ion-implanted MOSFETs with very small physical dimensions, *IEEE Journal of Solid State Circuits*, vol. SC-9, no. 5, pages 256–268, 1974. DOI: 10.1109/jssc.1974.1050511.

[27] Baccarani, G., Wordeman, M. R., and Dennard, R. H., Generalized scaling theory and its application to a 1/4 micrometer MOSFET design, *IEEE Transactions on Electron Devices*, vol. 31, no. 4, pages 452–462, 1984. DOI: 10.1109/t-ed.1984.21550.

[28] Haensch, W., Nowak, E. J., Dennard, R. H., Solomon, P. M., Bryant, A., Dokumaci, O. H., Kumar, A., Wang, X., Johnson, J. B., and Fischetti, M. V., Silicon CMOS devices beyond scaling, *IBM Journal of Research and Development*, vol. 50, no. 4/5, pages 339–361, 2006. DOI: 10.1147/rd.504.0339.

[29] Davari, B., Dennard, R. H., and Shahidi, G. G., CMOS scaling for high performance and low power-the next ten years, *Proc. of the IEEE*, vol. 83, no. 4, pages 595–606, 1995. DOI: 10.1109/5.371968.

[30] Sun, J.-Y. C., Taur, Y., Dennard, R. H., and Klepner, S. P., Submicrometer channel CMOS for low temperature operation, *IEEE Transactions on Electron Devices*, vol. 34, no. 1, pages 19–27, 1987. DOI: 10.1109/t-ed.1987.22881.

[31] Sai-Halasz, G. A., Wordeman, M. R., Kern, D. P., Ganin, E., Rishton, S., Zicherman, D. S., Schmid, H., Polcari, M. R., Ng, H. Y., Restle, P. J., Chang, T. H. P., and Dennard, R. H., Design and experimental technology for 0.1 μm gate-length low temperature operation FET's, *IEEE Electron Device Letters*, vol. 8, no. 10, pages 463–466, 1987. DOI: 10.1109/edl.1987.26695.

[32] Dennard, R. H., Cai, J., and Kumar, A., A perspective on today's scaling challenges and possible future directions, *Solid State Electronics*, vol. 51, no. 4, pages 518–525, 2007. DOI: 10.1016/j.sse.2007.02.004.

[33] Dennard, R. H., Gaensslen, F. H., Walker, E. J., and Cook, P. W., 1 μm MOSFET VLSI technology: Part II-device designs and characteristics for high-performance logic applications, *IEEE Transactions on Electron Devices*, vol. 26, no. 4, pages 325–333, 1979. DOI: 10.1109/t-ed.1979.19431.

[34] Dhar, S., Pattanaik, M., and Rajaram, P., Advancement in nanoscale CMOS device design en route to ultra low power applications, *VLSI Design*, pages 1–19, Hindwai Publishing Corporation, 2011. DOI: 10.1155/2011/178516.

[35] Kim, Y., Review paper: Challenges for nanoscale MOSFETs and emerging nanoelectronics, *Transactions on Electrical and Electronic Materials*, 10(1), 21, pages 23–41, 2009. DOI: 10.4313/teem.2010.11.3.093.

[36] Roy, K., Mukhopadhyay, S., and Mahmoodi, M. H., Leakage current mechanisms and leakage reduction techniques in deep submicrometer CMOS circuits, *Proc. of the IEEE*, vol. 91, no. 2, pages 305–327, 2003. DOI: 10.1109/jproc.2002.808156.

[37] Iwai, H., Roadmap for 22 nm and beyond, *Microelectronic Engineering*, vol. 86, no. 7–9, pages 1520–1528, 2009. DOI: 10.1016/j.mee.2009.03.129.

[38] Skotnicki, T., Hutchby, J. A., King, T.-J., Wong, H.-S. P., and Boeuf, F., The end of CMOS scaling: Toward the introduction of new materials and structural changes to improve MOSFET performance, *IEEE Circuits and Devices Magazine*, pages 16–26, 2005. DOI: 10.1109/MCD.2005.1388765.

[39] Narendra, S. G., Challenges and design choices in nanoscale CMOS, *ACM Journal on Emerging Technologies in Computing Systems*, vol. 1, no. 1, pages 7–49, 2005. DOI: 10.1145/1063803.1063805.

[40] Lin, S. and Banerjee, K., Cool chips: Opportunities and implications for power and thermal management, *IEEE Transactions on Electron Devices*, vol. 55, no. 1, pages 245–255, 2008. DOI: 10.1109/ted.2007.911763. 1, 2

[41] Shakouri, A. and Zhang, Y., On-chip solid state cooling for integrated circuits using thin-film microrefrigerators, *IEEE Transactions on Components and Packaging Technologies*, 28, pages 65–69, 2005. DOI: 10.1109/tcapt.2005.843219. 9

[42] Gaensslen, F. H., Rideout, V. L., Walker, E. J., and Walker, J. J., Very small MOSFETs for low temperature operation, *IEEE Transactions on Electron Devices*, vol. 24, no. 3, pages 218–229, 1977. DOI: 10.1109/t-ed.1977.18712. 1, 2, 9

[43] Semenov, O., Vassighi, A., and Sachdev, M., Impact of technology scaling on thermal behavior of leakage current in sub-quarter micron MOSFETs: Perspective of low temperature current testing, *Microelectronics Journal*, vol. 33, pages 985–994, 2002. DOI: 10.1016/s0026-2692(02)00071-x.

[44] Chang, L. and Hu, C., MOSFET scaling into the 10 nm regime, *Superlattices and Microstructures*, vol. 28, no. 5/6, pages 351–355, 2000. DOI: 10.1006/spmi.2000.0933.

[45] Kim, N. S., Austin, T., Blaauw, D., Mudge, T., Flautner, K., Hu, J. S., Irwin, M. J., Kandemir, M., and Narayanan, V., Leakage Current: Moore's law meets static power, computer, *IEEE Computers Society*, pages 68–75, 2003. DOI: 10.1109/mc.2003.1250885.

[46] Natori, K., Compact modeling of ballistic nanowire MOSFETs, *IEEE Transactions on Electron Devices*, vol. 55, no. 11, pages 2877–2885, 2008. DOI: 10.1109/ted.2008.2008009.

[47] Chang, L., Frank, D. J., Montoye, R. K., Koester, S. J., Ji, B. L., Cotens, P. W., Dennard, R. H., and Haensch, W., Practical strategies for power-efficient computing technologies, *Proc. of the IEEE*, vol. 98, no. 2, pages 215–236, 2010. DOI: 10.1109/jproc.2009.2035451.

[48] Keyes, R. W., Fundamental limits of silicon technology, *Proc. of the IEEE*, vol. 89, no. 3, pages 227–239, 2001. DOI: 10.1109/5.915372.

[49] Plummer, J. D. and Griffin, P. B., Materials and process limits in silicon VLSI technology, *Proc. of the IEEE*, vol. 89, no. 3, pages 240–257, 2001. DOI: 10.1109/5.915373.

[50] Wann, C. H., Noda, K., Tanaka, T., Yoshida, M., and Hu, C., A comparative study of advanced MOSFET concepts, *IEEE Transactions on Electron Devices*, vol. 43, no. 10, pages 1742–1753, 1996. DOI: 10.1109/16.536820.

[51] Ohmi, T., Sugawa, S., Kotari, K., Hirayma, M., and Morimoto, A., New paradigm of silicon technology, *Proc. of the IEEE*, vol. 89, no. 3, pages 394–412, 2001. DOI: 10.1109/5.915381.

[52] Markovic, D., Wang, C. C., Alarcon, L. P., Liu, T.-T., and Rabaey, J. M., Ultra low power design in near threshold region, *Proc. of the IEEE*, vol. 98, no. 2, pages 237–252, 2010. DOI: 10.1109/jproc.2009.2035453.

[53] Dreslinski, R. G., Wieckowski, M., Blaauw, D., Sylvester, D., and Mudge, T., Near-threshold computing: Reclaiming Moore's law through energy efficient integrated circuits, *Proc. of the IEEE*, vol. 98, no. 2, pages 253–266, 2010. DOI: 10.1109/jproc.2009.2034764.

[54] Gupta, S. K., Raychowdhury, A., and Roy, K., Digital computation in sub-threshold region for ultra-low power operation: A device-circuit-architecture code-sign perspective, *Proc. of the IEEE*, vol. 98, no. 2, pages 160–190, 2010. DOI: 10.1109/jproc.2009.2035060.

[55] Vitale, S. A., Wyatt, P. W., Checka, N., Kedzierski, J., and Keast, C. L., FDSOI process technology for subthreshold operation ultra-low power electronics, *Proc. of the IEEE*, vol. 98, no. 2, pages 333–342, 2010. DOI: 10.1149/1.3570794.

[56] Ronen, R., Mendelson, A., Lai, K., Lu, S.-L., Pollack, F., and Shen, J. P., Coming challenges in microarchitecture and architecture, *Proc. of the IEEE*, vol. 89, no. 3, pages 325–340, 2001. DOI: 10.1109/5.915377.

[57] Bryant, R. E., Cheng, K.-T., Kahng, A. B., Kentzer, K., Maly, W., Newton, R., Pileggi, L., Rabaey, J. M., and Vincentelli, A. S., Limitations and challenges of computer-aided design technology for CMOS VLSI, *Proc. of the IEEE*, vol. 89, no. 3, pages 341–365, 2001. DOI: 10.1109/5.915378.

[58] Doering, R. and Nishi, Y., Limits of integrated circuit manufacturing, *Proc. of the IEEE*, vol. 89, no. 3, pages 375–393, 2001. DOI: 10.1109/5.915380.

[59] Davis, J. A., Venketasan, R., Kaloyeros, A., Beylansky, M., Souri, S. J., Banerjee, K., Saraswat, K. C., Rahman, A., Reif, R., and Meindl, J. D., Interconnect limits on gigascale integration (GSI) in the 21st century, *Proc. of the IEEE*, vol. 89, no. 3, pages 305–324, 2001. DOI: 10.1109/5.915376.

[60] Yeo, Y.-C., King, T.-J., and Hu, C., MOSFET gate leakage current modeling and selection guide for alternative gate dielectrics based on leakage considerations, *IEEE Transactions on Electron Devices*, vol. 50, no. 4, pages 1027–1035, 2003. DOI: 10.1109/ted.2003.812504.

[61] Lu, W., Xie, P., and Lieber, C. M., Nanowire transistor performance limits and applications, *IEEE Transactions on Electron Devices*, vol. 55, no. 11, pages 2859–2876, 2008. DOI: 10.1109/ted.2008.2005158.

[62] Campera, A., Iannaccone, G., and Felice, C., Modeling of tunnelling currents in Hf-based gate stacks as a function of temperature and extraction of material parameters, *IEEE Transactions on Electron Devices*, vol. 54, no. 1, pages 83–89, 2007. DOI: 10.1109/ted.2006.887202.

[63] Narang, R., Saxena, M., Gupta, R. S., and Gupta, M., Impact of temperature variations on the device and circuit performance of tunnel FET: A simulation study, *IEEE Transactions on Nanotechnology*, vol. 12, no. 6, pages 951–957, 2013. DOI: 10.1109/tnano.2013.2276401.

[64] Datta, S., Liu, H., and Narayan, V., Tunnel FET technology: A reliability perspective, *Microelectronics Reliability*, vol. 54, pages 861–874, 2014. DOI: 10.1016/j.microrel.2014.02.002.

[65] Wilk, G. D., Wallace, R. M., and Anthony, J. M., High-k gate dielectrics: Current status and materials properties considerations, *Journal of Applied Physics*, vol. 89, no. 10, pages 5243–5275, 2001. DOI: 10.1063/1.1361065.

[66] Clark, R. D., Emerging applications for high K materials in VLSI technology, *Materials*, vol. 7, pages 2913–2944, 2014. DOI: 10.3390/ma7042913.

[67] Guha, S. and Narayan, V., High-K/metal gate science and technology, *Annual Review of Materials Research*, vol. 39, pages 181–202, 2009. DOI: 10.1146/annurev-matsci-082908-145320.

[68] Mishra, D., Iwai, H., and Wong, H., High-k gate dielectrics, interface, *The Electrochemical Society*, pages 30–34, 2005.

[69] Andro, T., Ultimate scaling of high-k gate dielectrics: Higher K or interfacial layer scavenging?, *Materials*, pages 478–500, 2012. DOI: 10.3390/ma5030478.

[70] Groner, M. D. and George, S. M., High-k dielectrics grown by atomic layer deposition: Capacitor and gate applications, Chapter 10, *Interlayer Dielectrics for Semiconductor Technologies*, pages 327–348, Elsevier Inc., 2003. DOI: 10.1016/b978-012511221-5/50012-x. 1, 2

[71] Takagi, S., Toriumi, A., Iwase, M., and Tango, H., On the universality of inversion layer mobility in Si MOSFET's: Part I-effects of substrate impurity concentration, *IEEE Transactions on Electron Devices*, vol. 41, no. 12, pages 2357–2362, 1994. DOI: 10.1109/16.337449. 5, 9

[72] Labounty, C., Shakouri, A., and Bowers, J. E., Design and characterization of thin-film microcoolers, *Journal of Applied Physics*, vol. 89, no. 7, pages 4059–4064, 2001. DOI: 10.1063/1.1353810. 7

[73] Wang, P., Cohen, B. A., and Yang, B., Analytical modeling of silicon thermoelectric microcooler, *Journal of Applied Physics*, vol. 100, no. 1, pages 014501–1 to 014501–13. DOI: 10.1063/1.2211328. 7

[74] Yang, M., Gusev, P. E., Ieong, M., Gluschenkov, O., Boyd, C. D., Chan, K. K., Kozlowski, M. P., D'Emic, P. C., Sicina, M. R., Jamison, C. P., and Chou, I. A., Performance dependence of CMOS on silicon substrate orientation for ultrathin oxynitride and HfO_2 gate dielectrics, *IEEE Electron Device Letters*, vol. 24, no. 5, pages 339–341, 2003. DOI: 10.1109/led.2003.812565. 1, 2, 9

[75] Mereu, B., Rossel, C., Gusev, E. P., and Yang, M., The role of Si orientation and temperature on the carrier mobility in metal oxide semiconductor field-effect transistors with ultrathin HfO_2 gate dielectrics, *Journal of Applied Physics*, vol. 100, no. 1, pages 014504–1 to 014504–6, 2006. DOI: 10.1063/1.2210627.

[76] Boucart, K. and Ionescu, M. A., Double-gate tunnel FET with high-k gate dielectric, *IEEE Transactions on Electron Devices*, vol. 54, no. 7, pages 1725–1733, 2007. DOI: 10.1109/ted.2007.899389. 51, 52

[77] Rustagi, C. S., Singh, N., Lim, Y. F., Zhang, G., Wang, S., Lo, G. Q., Balasubramanian, N., and Kwong, L. D., Low-temperature transport characteristics and quantum-confine effects in gate-all-around Si-nanowire N-MOSFET, *IEEE Electron Device Letters*, vol. 28, no. 10, pages 909–912, 2007. DOI: 10.1109/led.2007.904890. 55

[78] Leong, M., Doris, B., Kedzierski, J., Rim, K., and Yang, M., Silicon device scaling to the Sub-10-nm regime, *Science*, vol. 306, pages 2057–2060, 2004. DOI: 10.1126/science.1100731.

[79] Dabhi, K. C., Dasgupta, A., and Chauhan, S. Y., Computationally efficient analytical surface potential model for UTBB FD-SOI transistors, *IEEE International Conference on Emerging Electronics*, 2016. DOI: 10.1109/icemelec.2016.8074575.

[80] Pahwa, G., Dutta, T., Agarwal, A., Khandelwal, S., Salahuddin, S., Hu, C., and Chauhan, S. Y., Analysis and compact modeling of negative capacitance transistor with high ON-current and negative output differential resistance-part I: Model description, *IEEE Transactions on Electron Devices*, vol. 63, no. 12, pages 4981–4985, 2016. DOI: 10.1109/ted.2016.2614432.

[81] Pahwa, G., Dutta, T., Agarwal, A., Khandelwal, S., Salahuddin, S., Hu, C., and Chauhan, S. Y., Analysis and compact modeling of negative capacitance transistor with high ON-current and negative output differential resistance-part II: Model validation, *IEEE Transactions on Electron Devices*, vol. 63, no. 12, pages 4986–4992, 2016. DOI: 10.1109/ted.2016.2614436.

[82] Pahwa, G., Dutta, T., Agarwal, A., and Chauhan, S. Y., Compact model for ferroelectric negative capacitance transistor with MFIS structure, *IEEE Transactions on Electron Devices*, vol. 64, no. 3, pages 1366–1374, 2017. DOI: 10.1109/ted.2017.2654066.

[83] Campbell, J. P., Yu, L. C., Cheung, K. P., Qin, J., Suehle, J. S., Oates, A., and Sheng, K., Large random telegraph noise in sub-threshold operation of nano-scale n-MOSFETs, *IEEE International Conference on IC Design and Technology*, pages 17–20, 2009. DOI: 10.1109/icicdt.2009.5166255.

[84] Simoen, E., Kaczer, B., Luque, T. M., and Claeys, C., Random telegraph noise: From A device physicist's dream to a designer's nightmare, *ECS Transactions*, vol. 39, no. 1, pages 3–15, 2011. DOI: 10.1149/1.3615171.

[85] Tan, P. L. M., Lentaris, G., and Amaratunga, A. J. G., Device and circuit-level performance of carbon nanotube field-effect transistor with benchmarking against a nano-MOSFET, *Nanoscale Research Letters*, vol. 7(467), pages 1–10 , 2012. DOI: 10.1186/1556-276x-7-467.

[86] Sajjad, N. R., Chern, W., Hoyt, L. J., and Antoniadis, A. D., Trap assisted tunneling and its effect on subthreshold swing of tunnel FETs, *IEEE Transactions on Electron Devices*, vol. 63, no. 11, pages 4380–4387, 2016. DOI: 10.1109/ted.2016.2603468.

[87] Khaveh, T. R. H. and Mohammadi, S., Potential and drain current modeling of gate-all-around tunnel FETs considering the junctions depletion regions and the channel mobile charge carriers, *IEEE Transactions on Electron Devices*, vol. 63, no. 12, pages 5021–5029, 2016. DOI: 10.1109/ted.2016.2619761.

[88] Gupta, S., Ghosh, B., and Rahi, B. S., Compact analytical model of double-gate junction-less field effect transistor comprising quantum-mechanical effect, *Journal of Semiconductors*, vol. 36, no. 2, pages 02400–1 to 024001–6, 2015. DOI: 10.1088/1674-4926/36/2/024001.

[89] Panchore, M., Singh, J., and Mohanty, P. S., Impact of channel hot carrier effect in junction and doping-free devices and circuits, *IEEE Transactions on Electron Devices*, vol. 63, no. 12, pages 5068–5071, 2016. DOI: 10.1109/ted.2016.2619621.

[90] Timp, G., Bude, J., Baumann, F., Bourdelle, K. K., Boone, T., Garno, J., Ghetti, A., Green, M., Gossmann, H., Kim, Y., Kleiman, R., Kornblit, A., Klemens, F., Moccio, S., Muller, D., Rosamilia, J., Silverman, P., Sorsch, T., Timp, W., Tennant, D., Tung, R., and Weir, B., The relentless march of the MOSFET gate oxide thickness to zero, *Microelectronics Reliability*, vol. 40, pages 557–562, 2000. DOI: 10.1016/s0026-2714(99)00257-7.

[91] Reggiani, S., Valdinoci, M., Colalongo, L., Rudan, M., and Baccarani, G., An analytical temperature-dependent model for majority- and minority-carrier mobility in silicon devices, *VLSI Design*, vol. 10, no. 4, pages 467–483, 2000. DOI: 10.1155/2000/52147. 32

[92] Agopian, D. G. P., Martino, V. D. M., Filho, S. D. G. S., Martino, A. J., Rooyackers, R., Leonelli, D., and Claeys, C., Temperature impact on the tunnel FET off state current components, *Solid State Electronics*, vol. 78, pages 141–146, 2012. DOI: 10.1016/j.sse.2012.05.053.

[93] Huang, Z. J., Long, P., Povolotskyi, M., Rodwell, J. W. M., and Klimeck, G., Exploring channel doping designs for high-performance tunneling FETs, *74th Annual Device Research Conference*, pages 1–2, 2016. DOI: 10.1109/drc.2016.7548456.

[94] Long, P., Povolotskyi, M., Huang, Z. J., Illatikhameneh, H., Ameen, T., Rahman, R., Kubis, T., Klimeck, G., and Rodwell, J. W. M., Extremely high simulated ballistic currents in triple-heterojunction tunnel transistors, *74th Annual Device Research Conference*, pages 1–2, 2016. DOI: 10.1109/drc.2016.7548424.

[95] Matheu, P., *Investigations of Tunneling For Field Effect Transistors*, pages 1–79, Doctoral Thesis, Applied Science and Technology in the Graduate Division of University of California at Berkeley, 2012.

[96] Lochtefeld, A., Djomehri, I. J., Samudra, G., and Antoniadis, D. A., New insights into carrier transport in n-MOSFETs, *IBM Journal Research and Development*, vol. 46, no. 2/3, pages 347–357, 2002. DOI: 10.1147/rd.462.0347.

[97] Groner, M. D. and George, S. M., High-k dielectrics grown by atomic layer deposition: Capacitor and gate applications, Chapter 10, *Interlayer Dielectrics for Semiconductor Technologies*, pages 327–348, Elsevier Inc., 2003. DOI: 10.1016/b978-012511221-5/50012-x.

[98] Lu, W., Xie, P., and Lieber, C. M., Nanowire transistor performance limits and applications, *IEEE Transactions on Electron Devices*, vol. 55, no. 11, pages 2859–2876, 2008. DOI: 10.1109/ted.2008.2005158. 52, 53, 54

[99] Manut, B. A., Zhang, J. F., Duan, M., Ji Z., Zhang, W. D., Kaczer, B., Schram, T., Horiguchi, N., and Groeseneken, G., Impact of hot carrier aging on random telegraph noise and within a device fluctuation, vol. 4, no. 1, pages 15–21, 2016. DOI: 10.1109/jeds.2015.2502760.

[100] Thean, A., Challenges and enablers of logic CMOS scaling in the next 10 years, Presentation *IMEC Technology Forum*, slides 1–36, Taiwan, 2013.

[101] Moselund, K. E., Nazmjadeh, M., Dobrosz, P., Olsen, H. S., Bouvet, D., Michielis, D. L., Pott, V., and Ionescu, M. A., The high-mobility bended n-channel silicon nanowire transistor, *IEEE Transactions on Electron Devices*, vol. 57, no. 4, pages 866–876, 2010. DOI: 10.1109/ted.2010.2040939.

[102] Khan, I. A., Bhowmik, D., Yu, P., Kim, S. J., Pan, X., Ramesh, R., and Salahuddin, S., Experimental evidence of ferroelectric negative capacitance in nanoscale heterostructures, *Applied Physics Letters*, vol. 99, pages 113501–1–3, 2011. DOI: 10.1063/1.3634072. 58

[103] Salvatore, A. G., Rusu, A., and Ionescu, M. A., Experimental confirmation of temperature dependent negative capacitance in ferroelectric field effect transistors, *Applied Physics Letters*, vol. 100, pages 163504–1–4, 2012. DOI: 10.1063/1.4704179. 58

[104] Kobayashi, M. and Hiramoto, T., On device design for steep-slope negative capacitance field-effect transistor operating at sub-0.2 V supply voltage with ferroelectric HfO_2 thin film, *AIP Advances*, vol. 6, pages 025113–1–10, 2016. DOI: 10.1063/1.4942427.

[105] Liu, F, Zhou, Y., Wang, Y. Liu, X., Wang, J., and Guo, H., Negative capacitance transistors with monolayer black phosphorous, *Nature Publication Journal (Quantum Materials)*, vol. 1, pages 16004–1–6, 2016. DOI: 10.1038/npjquantmats.2016.4.

[106] Akkez, B. I., Beranger, F. C., Cros, A., Balestra, F., and Ghibuado, G., Evidence of mobility enhacement due to back biasing in UTBOX FDSOI high-k metal gate technology, *IEEE SOI-3D-Subthreshold Microelectronics Technology Unified Conference*, pages 1–2, 2013. DOI: 10.1109/s3s.2013.6716551.

[107] Coquand, R., Jaud, A.-M., Rozeau, O., ElOudrhiri, I. A., Martinie, S., Triozon, F., Pons, N., Barruad, S., Monfray, S., Boeuf, F., Ghibaudo, G., and Faynot, O., Comparative simulation of TriGate and FinFET on SOI: Evaluating a multiple threshold voltage strategy on triple gate devices, *IEEE SOI-3D-Subthreshold Microelectronics Technology Unified Conference*, pages 1–2, 2013. DOI: 10.1109/s3s.2013.6716523.

[108] Zhang, J., Trommer, J., Weber, M. W., Gaillardon, E.-P., and Micheli, D. G., On temperature dependency of steep subthreshold slope in dual-independent-gate FinFET, *IEEE Journal of the Electron Devices Society*, vol. 3, no. 6, pages 452–456, 2015. DOI: 10.1109/jeds.2015.2482123.

[109] Li, M. O., Esseni, D., Nahas, J. J., Jena, D., and Xing, G. H., Two-dimensional heterojunction interlayer tunneling field effect transistors (thin T-FETs), *IEEE Journal of the Electron Devices Society*, vol. 3, no. 3, pages 200–207, 2015. DOI: 10.1109/jeds.2015.2390643.

[110] Lu, H. and Seabaugh, A., Tunnel field-effect transistors: State-of-the-art, *IEEE Journal of the Electron Devices Society*, vol. 2, no. 4, pages 44–49, 2014. DOI: 10.1109/jeds.2014.2326622.

[111] Avci, U.-E., Morris, H. D., and Young, A. I., Tunnel field-effect transistors: Prospects and challenges, *IEEE Journal of the Electron Devices Society*, vol. 3, no. 3, pages 88–95. DOI: 10.1109/jeds.2015.2390591.

[112] Zhao, T-Q., Richter, S., Braucks, S. C., Knoll, L., Blaeser, S., Luong, V. G., Trellenkamp, S., Schäfer, A., Tiedemann, A., Hartmann, M.-J., Bourdelle, K., and Mantl, S., Strained Si and SiGe nanowire tunnel FETs for logic and analog applications, *IEEE Journal of the Electron Devices Society*, vol. 3, no. 3, pages 103–114, 2015. DOI: 10.1109/jeds.2015.2400371. 1, 2, 9

[113] Arora, D. N., Hauser, R. J., and Roulston, J. D., Electron and hole mobilities in silicon as a function of concentration and temperature, *IEEE Transactions of Electron Devices*, vol. 29, no. 2, pages 292–295, 1982. DOI: 10.1109/t-ed.1982.20698. 5

[114] Ionescu, M. A. and Riel, H., Tunnel field-effect transistors as energy-efficient electronic switches, *Nature*, vol. 479, pages 329–337, 2011. DOI: 10.1038/nature10679. 49, 50

[115] Cui, Y., Zhong, Z., Wang, D., Wang, U. W., and Lieber, M. C., High performance silicon nanowire field effect transistors, *Nano Letters*, vol. 3, no. 2, pages 149–152, 2003. DOI: 10.1021/nl0258751. 53

[116] Sahay, S. and Kumar, J. M., Physical insights into the nature of gate-induced drain leakage in ultrashort channel nanowire FETs, *IEEE Transactions on Electron Devices*, vol. 64, no. 6, pages 2604–2610, 2017. DOI: 10.1109/ted.2017.2688134. 55

[117] Lee, H., Yoon, Y., and Shin, C., Current-voltage for negative capacitance field-effect transistors, *IEEE Electron Device Letters*, vol. 38, no. 5, pages 669–672, 2017. DOI: 10.1109/led.2017.2679102. 56, 57

[118] Khandelwal, S., Duarte, P. J., Khan, I. A., Salahuddin, S., and Hu, C., Impact of parasitic capacitance and ferroelectric parameters on negative capacitance FinFET characteristics, *IEEE Electron Device Letters*, vol. 38, no. 1, pages 142–144, 2017. DOI: 10.1109/led.2016.2628349. 58

[119] Ko, E., Lee, W. J., and Shin, C., Negative capacitance FinFET with sub-20-mV/decade subthreshold slope and minimal hysteresis of 0.48 V, *IEEE Electron Device Letters*, vol. 38, no. 4, pages 418–421, 2017. DOI: 10.1109/led.2017.2672967. 58

Author's Biography

NABIL SHOVON ASHRAF

Dr. Nabil Shovon Ashraf was born in Dhaka, Bangladesh in 1974. Currently, Dr. Ashraf serves as an Associate Professor in the Department of Electrical and Computer Engineering of North South University, Dhaka, Bangladesh where he had previously served as an Assistant Professor from September 2014–June 2018. He obtained a Bachelor of Technology degree in Electrical Engineering from Indian Institute of Technology Kanpur, India in 1997. He obtained a Master of Science degree in Electrical Engineering from University of Central Florida, Orlando, USA in 1999. In 2011, he obtained his Ph.D. in Electrical Engineering from Arizona State University, Tempe, USA. From December 2011–May 2014, he was a Post Doctoral Researcher in the department of Electrical Engineering of Arizona State University Tempe. He was employed as design engineer in RF Monolithics Inc., a surface acoustic wave () based filter design company in Dallas, Texas, USA from August 1999–March 2001. From October 2003–June 2006, he served on the faculty as Assistant Professor of the Department of Electrical and Electronic Engineering at the Islamic University of Technology, Gazipur, Bangladesh. To date Dr. Ashraf has published 6 peer-reviewed journal articles (two IEEE EDS) and 15 international conference proceedings (3 IEEE EDS). In 2016, he published *New Prospects of Integrating Low Substrate Temperatures with Scaling-Sustained Device Architectural Innovation* with Morgan & Claypool and contributed to two book chapters on interface trap-induced threshold voltage fluctuations in the presence of random channel dopants of scaled n-MOSFET at the invitations of highly accomplished international book editors. He was cited in *Who's Who in America for Marquis* online biographies of distinguished and eminent researchers for two consecutive years—2015 (69th edition) and 2016 (70th platinum edition). In 2017, Dr. Ashraf became the recipient of the Albert Nelson Marquis Lifetime Achievement Award honored by Marquis Who's Who. He specializes in the area of device physics and modeling analysis of scaled devices for enabling improved device performance at the scaled node of current MOSFET device architectures.